JOURNEYMAN ELECTRICIAN'S

Exam Preparation

Based on the

2020 NEC® National Electrical Code

atp AMERICAN TECHNICAL PUBLISHERS
ORLAND PARK, ILLINOIS

R. E. Chellew

Journeyman Electrician's Exam Preparation Based on the 2020 NEC® contains procedures commonly practiced in industry and the trade. Specific procedures vary with each task and must be performed by a qualified person. For maximum safety, always refer to specific manufacturer recommendations, insurance regulations, specific job site and plant procedures, applicable federal, state, and local regulations, and any authority having jurisdiction. The material contained herein is intended to be an educational resource for the user. American Technical Publishers assumes no responsibility or liability in connection with this material or its use by any individual or organization.

American Technical Publishers, Inc., Editorial Staff

Editor in Chief:
Peter A. Zurlis
Assistant Production Manager:
Nicole D. Burian
Technical Editor:
William A. Hendricks
Supervising Copy Editor:
Catherine A. Mini
Copy Editor:
Dane K. Hamann
Art Manager:
Sarah E. Kaducak

Cover Design:
Steven E. Gibbs
Cover Photos:
Christopher T. Proctor
Illustration/Layout:
Richard O. Davis
Nicholas G. Doornbos
Nicholas W. Basham
Digital Media Manager
Adam T. Schuldt
Digital Resources:
Robert E. Stickley
Lauren M. Lenoir

Acknowledgments

The author and publisher are grateful to the Pinellas County Construction Licensing Board for providing information and technical assistance.

The author and publisher appreciate the technical review provided by Clyde J. Hady, Coordinator, Bay Area Electrical Apprenticeship Program, North Clearwater, Florida.

Journeyman Electrician's Exam Preparation Based on the 2020 NEC® contains technical and electrical terms and definitions that have been derived for educational purposes from *NFPA 70®*, *National Electrical Code®* (NEC®), 2020 edition.

1 2 3 4 5 6 7 8 9 – 20 – 9 8 7 6 5 4 3 2 1

Printed in the United States of America

ISBN 978-0-8269-2041-6

This book is printed on recycled paper.

Contents

CHAPTER

3 NEC® Examples 37

Conduit Fill • Conductor Adjustment Factor • Temperature Correction • Single Motors 1ϕ • Full Load Current • Conductor Size • Overloads • Overload Protection • Disconnecting Means • Motor Feeders • Ranges (Household) • Calculating Box Fill for Conductors • Conductor Resistance • Derating Conductors Installed on Rooftops • Electric Welders • Service Calculations • Service Load Calculations

CHAPTER

4 Journeyman Electrician's Practice Licensing Exams (45 minutes) 59

CHAPTER

5 Journeyman Electrician's Practice Licensing Exams (180 minutes) 93

Online Learner Resources

- Quick Quizzes®
- Illustrated Glossary
- Flash Cards
- Voltage Drop Calculator
- Sample Licensing Exams
- Media Library
- Internet Resources

Book Features

Journeyman Electrician's Exam Preparation Based on the 2020 NEC® is designed to help individuals prepare for the journeyman electrician's electrical licensing examinations. This workbook is based on the 2020 National Electrical Code® (NEC®) and assumes that the applicant has a basic understanding of electrical theory and trade knowledge. The review questions and practice licensing exams included are similar to the questions on a state or local electrical licensing exam. Answers to the review questions and practice licensing exams are listed in the answer key located in the back of this workbook. *Journeyman Electrician's Exam Preparation Based on the 2020 NEC®* also includes over 500 additional sample licensing exam questions as part of the complimentary online learner resources.

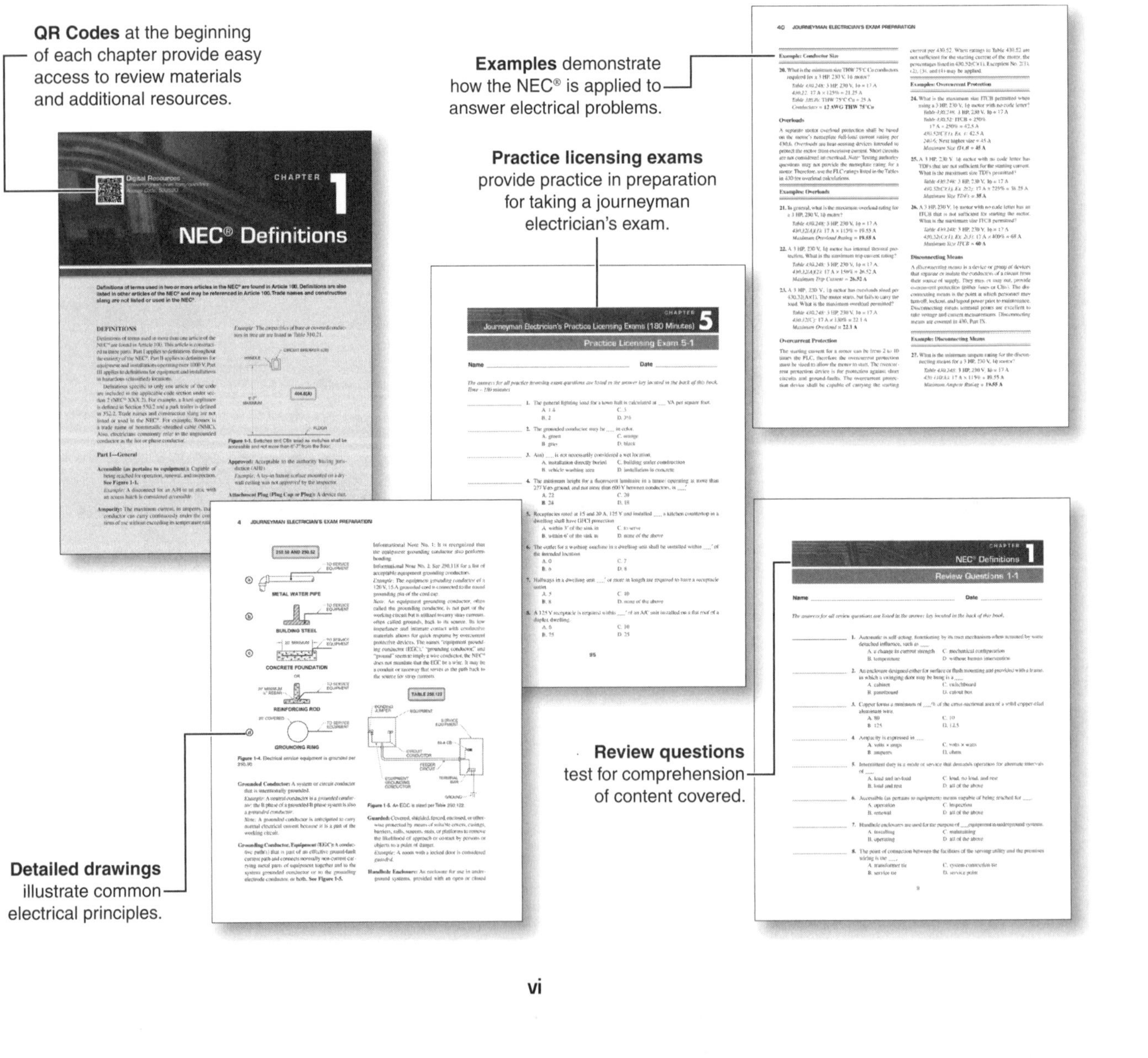

The Sample Licensing Exams in the online learner resources provide interactive questions designed to help applicants prepare for the journeyman electrician's exam. These sample exams test comprehension of electrical theory, trade knowledge, and the NEC®.

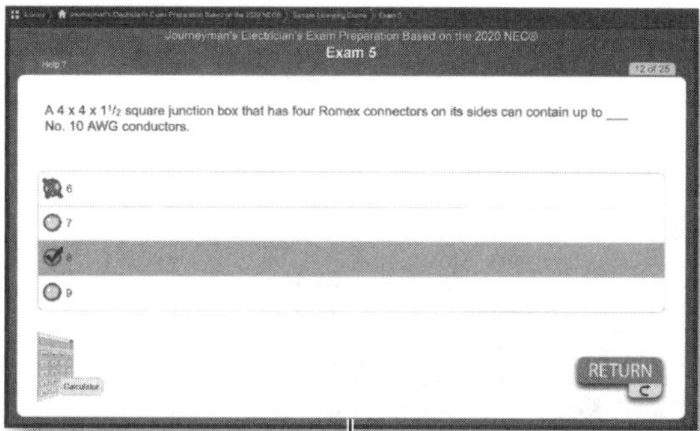

Over 500 additional questions have been added to the Sample Licensing Exams provided as part of the online learner resources.

Both *Journeyman Electrician's Exam Preparation Based on the 2020 NEC®* and the 2020 NEC® Codebook may be purchased together from ATP for a discounted price.

	Item No.	List Price
Workbook with NEC® Codebook	2041NEC	165.25

National Electrical Code® 2020

by National Fire Protection Association®
2020, softcover, 901 pages.

The *National Electrical Code®* is the benchmark standard for electrical safety in residential, commercial, and industrial settings. The 2020 edition provides the latest requirements for safer and more effective electrical design, installation, and inspection, including provisions for wiring, overcurrent protection, grounding, and equipment.

Online Learner Resources

Journeyman Electrician's Exam Preparation Based on the 2020 NEC® includes access to online learner resources that supplement workbook content and better prepare applicants for the licensing exam. These online resources can be accessed using either of the following methods:

- Key atplearningresources.com/quicklinks into a web browser and enter QuickLinks™ access code **533920**.
- Use a Quick Response (QR) reader app to scan the QR Code with a mobile device.

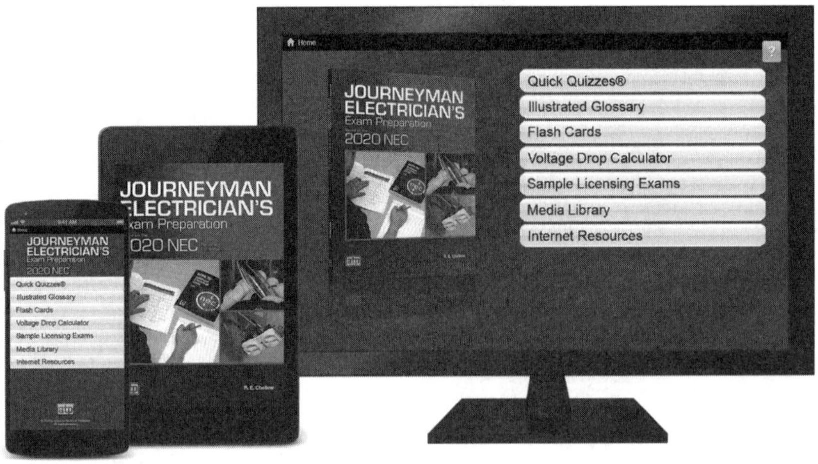

The online learner resources include the following:

- **Quick Quizzes®** that provide interactive questions for each chapter, with embedded links to highlighted content within the workbook and to the Illustrated Glossary
- **Illustrated Glossary** that serves as a helpful reference to commonly used terms, with selected terms linked to workbook illustrations
- **Flash Cards** that provide a self-study/review of common terms and their definitions
- **Voltage Drop Calculator** that serves as a useful tool for calculating voltage drop for various types of circuits
- **Sample Licensing Exams** that provide practice in preparing for a journeyman electrician's exam, each with a timer to help simulate an examination setting
- **Media Library** that consists of videos and animations that reinforce workbook content
- **Internet Resources** that provide access to additional online resources to support continued learning

Preparing to Take the Journeyman Electrician's Exam

The type of work and duties of a journeyman electrician are generally determined by local and state jurisdiction. The journeyman electrician usually performs electrical work under the supervision of a master electrician.

A journeyman electrician's exam consists of questions and problems prepared and administered by a local licensing board. Most local exams are given periodically at a time and place specified by the licensing board. Some licensing boards administer the exam in parts, which are graded separately. Request an exam breakdown when making an application.

The licensing board establishes requirements of eligibility of applicants for a journeyman electrician's exam. The applicant is usually required to show four years experience in the electrical trade. Eligible applicants are administered an exam to determine if they have the knowledge to perform electrical work on the journeyman level. The quality of their work must meet the requirements of local, state, and national codes. It must be a safe electrical installation.

Applicants who pass a journeyman electrician's exam are issued a journeyman card. The card contains the name of the licensing board, the name of the cardholder, license number, expiration date, and issuance date. **See Figure 1.**

Application and Exam Fees

Exam form cover letters contain information regarding journeyman electrician's exam forms and general exam information. Exam forms are usually provided by the licensing board. **See Figure 2.**

Forms detailing guidelines for experience may also be provided by the licensing board. Generally, four years of work experience in the trade, under the direction of an electrical contractor or master electrician, is required for eligibility to take the journeyman electrician's exam. Partial credit may be given for military service, school courses, and other related work experience. **See Figure 3.**

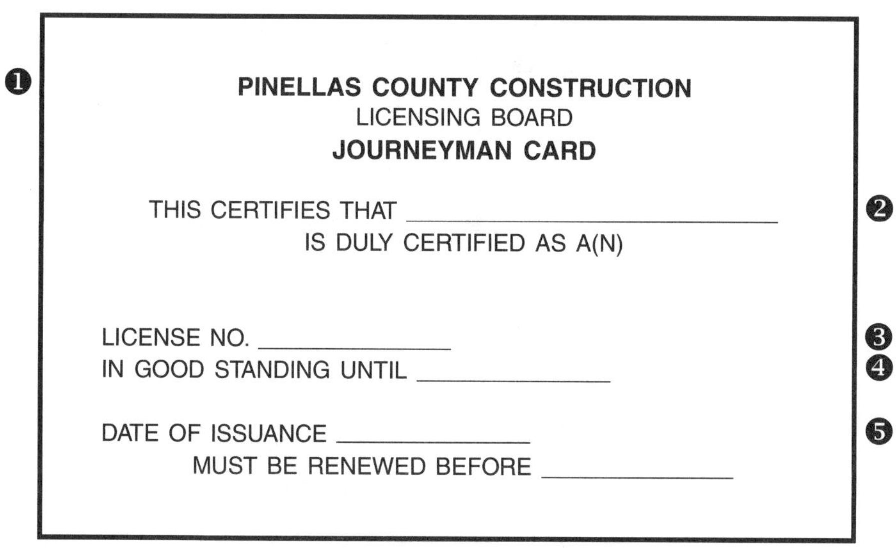
❶ PINELLAS COUNTY CONSTRUCTION
LICENSING BOARD
JOURNEYMAN CARD
THIS CERTIFIES THAT ______________ ❷
IS DULY CERTIFIED AS A(N)
LICENSE NO. ______________ ❸
IN GOOD STANDING UNTIL ______________ ❹
DATE OF ISSUANCE ______________ ❺
MUST BE RENEWED BEFORE ______________

❶ AUTHORITY HAVING JURISDICTION
❷ CARDHOLDER
❸ LICENSE NUMBER
❹ EXPIRATION DATE
❺ ISSUANCE DATE

Pinellas County Construction Licensing Board

Figure 1. Applicants who pass the journeyman electrician's exam are issued a journeyman card by the authority having jurisdiction (AHJ) in the area.

PINELLAS COUNTY
CONSTRUCTION LICENSING BOARD

TO: APPLICANTS FOR JOURNEYMAN EXAMINATION

REF: CHAPTER 75-489, LAWS OF FLORIDA

Each applicant must complete the attached application forms with supporting documents and file them with the Board. The Committees meet the first Tuesday and Wednesday of each month.

The next committee meeting is ______________________________

Your application should be submitted not later than ______________________________

Applications must be signed and notarized as indicated. ❹

❶ Examinations are open-book and prepared for the PCCLB by an independent testing agency. Questions are multiple-choice, fill-in, or true-false.

❷ All books and references must be brought by the examinee. No class notes, scratch-paper, exam study aids or like materials will be allowed. Silent, battery-powered, non-printing calculators are permitted, and applicants must bring their own pencils (#2), pens and rulers.

❸ Most examinations take about six hours with a lunch break of approximately one hour. Smoking is prohibited in the examination area. A passing grade is 70%. You will be notified of your grade within 30 days. ❺

<u>Do not call the PCCLB office before 30 days for your examination results.</u>

If an applicant for an original certificate does not provide a complete application with all required supplementary information within one year from the date of filing the application, the fee paid shall be credited to the PCCLB as an earned fee. A new application for a certificate shall be accompanied by another application fee.

All experience requirements must be verified on attachment B. The form may be reproduced as needed and all verification must be notarized.

Application Fee .... $75.00 ❻

<u>ALL FEES ARE NON-REFUNDABLE</u>

7887 Bryan Dairy Road, Suite 133 • Largo, Florida 33777 • Phone 727-582-3100 • www.pcclb.com

Pinellas County Construction Licensing Board

❶ **TYPE OF EXAM**

❷ **MATERIAL THAT MAY BE BROUGHT**

❸ **LENGTH OF EXAM**

❹ **SIGNATURE AND NOTARIZATION REQUIRED**

❺ **MINIMUM PASSING GRADE**

❻ **COST OF EXAM**

Figure 2. A journeyman electrician's exam cover letter contains general information regarding the exam.

PINELLAS COUNTY
CONSTRUCTION LICENSING BOARD

TO PERSONS REQUESTED TO CERTIFY TO APPLICANT'S EXPERIENCE:

The applicant named is required to prove his right to take a journeyman's examination by furnishing these certificates in support of his/her experience shown in the application. Enough verification is required to prove such experience to the Board. The applicant must have had:

Not less than four (4) years' experience preceding the filing of an application in the particular classification of license for which application is made:	• ELECTRICIANS • PLUMBERS • PIPE FITTERS • AIR CONDITIONING
Not less than three (3) years' experience preceding the filing of an application in the particular classification of license for which application is made:	• SHEET METAL

You are being requested to certify as to your knowledge of applicant experience by completing the attached form. Verifications must be by **licensed contractors** with direct knowledge of the applicant's experience.

THE FORM MUST BE NOTARIZED.

DO NOT MAIL THE FORM TO THE PINELLAS COUNTY CONSTRUCTION LICENSING BOARD. RETURN IT TO THE APPLICANT IN ORDER FOR HIM/HER TO ATTACH TO HIS/HER APPLICATION.

Your cooperation is earnestly solicited so that the PCCLB can determine whether an applicant has had the experience necessary to become a capable and qualified journeyman.

HOW TO COMPUTE YEARS OF EXPERIENCE – USE ANY OF THE FOLLOWING:

❶ 1. Completion of a 4-year apprenticeship program equals 4 years; or

❷ 2. Completion of 4 years' full-time employment (2,000 hours/year) since leaving (high) school equals 4 years; or

❸ 3. Completion of less than 4 years' full-time employment since leaving (high) school can be added to actual part-time hours worked before leaving school once age 16 was attained. This means actual part-time hours worked since a 16th birthday and during school years may be added to years of work since leaving school to total 4 years.
The total hours needed to qualify for 4 years equals 8,000 (4 x 2,000). For the sheet metal journeyman classification only verification of 6,000 hours is required.
All work must be verified by licensed contractors or building officials familiar with the applicant's experience.

7887 Bryan Dairy Road, Suite 133 • Largo, Florida 33777 • Phone 727-582-3100 • www.pcclb.com

Pinellas County Construction Licensing Board

❶ **COMPLETION OF A 4-YEAR APPRENTICESHIP PROGRAM**

❷ **4 YEARS FULL-TIME EMPLOYMENT IN TRADE**

❸ **PART-TIME WORK HOURS STARTING AT AGE 16 MAY BE ADDED TO FULL TIME EMPLOYMENT TO COMPLETE 8000 HOURS**

Figure 3. Four years of work experience in the trade is generally required to take a journeyman electrician's exam.

Information in such forms may include:

- Minimum years of trade experience required
- Conditions of trade experience
- Partial credit

The journeyman application form contains spaces for the applicant to provide the information required by the licensing board. The licensing board requires that the applicant complete the journeyman application form and pay an exam fee. The information required will vary somewhat from jurisdiction to jurisdiction. **See Figure 4.** Generally, however, the following is required:

- Type of certificate applied for
- Personal information (address, telephone, etc.)
- Schools attended
- Employment
- Letters of reference
- Signature and notarization

Certificates in support of journeyman applicant's experience qualifications may also be required. **See Figure 5.** Such certifications must be made by former employers licensed as contractors. The certificate contains:

- Dates of employment
- Statement of applicant's experience
- Signature of employer
- Notarization

The local licensing board generally accepts copies of an applicant's IRS W-2 forms to verify work experience. The W-2 form may be used in cases where an applicant's ex-employer is no longer in business or cannot be located.

Acceptance

Upon fulfilling the requirements of the licensing board, an applicant will receive confirmation of an exam date. This includes the location and time of the exam. A journeyman electrician's exam is generally given once a month. Jurisdictions may vary. Confirm availability for exam dates with the local licensing board.

Reference Books

The local licensing board usually provides a list of books and study aids permitted in the exam room. Books and study aids other than those listed should be approved by the licensing board or the testing proctor.

Amendments

The local licensing board may have amendments to the National Electrical Code® (NEC®). The board may add or delete sections to the NEC®. A local test may incorporate local amendments. When filling out an application, request a copy of amendments and ask if they are included in the exam.

Private Testing Agencies

The local licensing board may make up and administer their own exams or use a private testing agency. The licensing board sets the criteria for the exam. The private agency administers the exam and returns the results to the licensing board.

Special Exams

The need may arise to have a license as quickly as possible. The local licensing board may permit a special exam. At the time of application, request this information, if it is necessary.

Schools/Bookstores

The local licensing board may supply the applicant with a list of public schools, private schools, and bookstores where exam preparation classes are conducted, or where technical material is available for purchase. Ask for this information, if it applies.

Exam Preparation

1. Be prepared, mentally ready, and know and understand the basic material that is covered on the exam.
2. Only conduct a light review the night before the exam.
3. Get a good night's sleep.
4. Eat a light breakfast.
5. Be relaxed, alert, and free of tension.
6. Allow sufficient travel time to get to the exam site.
7. Determine the amount of questions per exam section. Calculate the time per question and divide the sections into portions of time. This allows an applicant to pace himself or herself. Generally, questions not answered are marked incorrect.
8. Read each question carefully. Fully understand the question before marking the answer sheet.
9. Time is critical. Too much time spent on any one question could result in questions not answered.
10. Use the Dash Method. When an applicant has used his or her allotted time for a question, mark a light dash on the answer key alongside the question number. Proceed to the next question. Upon exam completion, go back and answer the dashed questions that are not complete.
11. Generally, if an applicant guesses, the first choice is usually right. An answer should not be changed unless it is certain that it is wrong.
12. Tab the NEC® book. Tabs are available at any book store or department store.

PINELLAS COUNTY
CONSTRUCTION LICENSING BOARD

JOURNEYMAN APPLICATION FORM

Cert. No. ____________ Receipt. No. ____________ Date: ____________

DATE OF EXAM	GRADE
____________	____________
____________	____________
____________	____________

DO NOT FILL IN ABOVE LINE

Type or Print in ink. Answer all questions.

❶ 1. Present Journeyman Certificate No. ____________

❷ 2. I am applying for the following:
- ☐ Plumbing Journeyman
- ☒ Electrical Journeyman
- ☐ Sheet Metal Journeyman
- ☐ Pipe Fitting Journeyman
- ☐ Air Conditioning Journeyman

3. Have you previously applied to this Board for licensure as a contractor or journeyman? If so, when ? ____________

4. Name of individual to be certified (as you would have it on Certificate) ____________

5. Residence address ____________ City ____________ Zip ____________

6. Date of birth ____________ Place of birth ____________

7. Gender ☐ Male ☐ Female Social Security Number ____________
(The PCCLB collects your social security number for the following purposes: classification of accounts; identification and verification; credit worthiness; billing and payments; data collection, reconciliation, tracking, benefit processing, and tax reporting. Social security numbers are also used as a unique numeric identifier and may be used for research purposes).

8. Telephone Home ____________ Telephone Business ____________

9. Education: Give names of institutions, locations, length of time spent in each course:

❸

	School Name	Location	Years Attended	Type of Course
High School	____________	____________	From ______ To ______	____________
Trade School	____________	____________	From ______ To ______	____________
Other	____________	____________	From ______ To ______	____________

7887 Bryan Dairy Road, Suite 133 • Largo, Florida 33777 • Phone 727-582-3100 • www.pcclb.com

Pinellas County Construction Licensing Board

❶ **TYPE OF CERTIFICATE APPLIED FOR**

❷ **PERSONAL INFORMATION**

❸ **SCHOOLS ATTENDED**

Figure 4. *(continued on next page)*

10. Have you previously been employed as any type of licensed journeyman?

How long? ______________ Type of Journeyman ______________

Where? ______________

11. History of employment as a helper, apprentice or journeyman in the profession for which you desire certification: (use last six years only).

❹

DATE: MONTH/YEAR		NAME OF FIRM OR JOB	WHERE	DUTIES
FROM	TO			

❺ 12. Three original character letters must be submitted from reputable business or professional persons (not former employers or relatives of the applicant) of Pinellas County or the county of applicant's last business location.

13. If the answer to any of the following questions is "yes," explain fully under "Remarks," giving names, dates, locations, etc.

	Yes	No
a. Have you been convicted of a misdemeanor or felony, or are you presently under a charge of committing a misdemeanor or felony?	____	____
b. Have you ever been refused a journeyman's or other professional license or had such a license suspended or revoked?	____	____
c. Are you now doing business, or have you ever done business, under a fictitious name: Attach proof of compliance with the Florida Fictitious Name Act.	____	____

Remarks: ______________

14. I hereby apply for a Journeyman's Certificate as a(n) ______________ Journeyman and enclose the fee in the amount of $75.00. I have read the application and the accompanying instruction sheet and have answered all questions. I understand that my certificate can be suspended or revoked for good cause shown.

STATE OF FLORIDA
COUNTY OF ______________

____
Applicant Signature

❻ The foregoing instrument was acknowledged before me

This ____ day of ______________, 20____ by:

____
Date

____
Name of person taking acknowledgement

____
Signature of Notary

____
Print, Type or Stamp Commissioned Name of Notary Public

Personally known ____ or Produced Identification ____

Type of Identification Produced ______________

7887 Bryan Dairy Road, Suite 133 • Largo, Florida 33777 • Phone 727-582-3100 • www.pcclb.com

Pinellas County Construction Licensing Board

❹ **EMPLOYMENT**

❺ **LETTERS OF REFERENCE**

❻ **SIGNATURE AND NOTARIZATION**

Figure 4. The journeyman application form provides personal information.

PINELLAS COUNTY CONSTRUCTION LICENSING BOARD

Information in this box to be filled in by Applicant.
(PRINT OR TYPE)

Applicant ______________________
(Name of person to be examined)

Address ______________________
(Same as application)

Classification Journeyman Electrician

CERTIFICATE IN SUPPORT OF JOURNEYMAN APPLICANT'S EXPERIENCE QUALIFICATIONS
(To be attached to application for examination)

The person certifying to his knowledge of the experience of the applicant named above shall complete the form below.
READ THE INSTRUCTIONS BEFORE PROCEEDING.

I, ______________________ certify that I am
Name of Employer (PRINT)

personally familiar with the work experience of ______________________ ❶
Name of Examinee

during the period from ______________ to ______________
Month Year Month Year

and that I know of my own direct knowledge that said applicant was employed as follows:

(Tell in your own words what you know of applicant's experience. Give the dates of employment. Describe the type of work performed and the position as worker, apprentice, helper, journeyman, foreman, or supervisory employee. Describe the kind of buildings, structures, projects or equipment worked on and the extent of trade experience known to you. Give any other details that might aid in evaluating experience.) ❷

STATE OF FLORIDA
COUNTY OF ______________

The foregoing instrument was acknowledged before me

This ____ day of ______________, 20____ by:

____
Name of person taking acknowledgement

❹ ______________________
Signature of Notary

____
Print, Type or Stamp Commissioned Name of Notary Public

Personally known ____ or Produced Identification ____

Type of Identification Produced ______________

______________________ ❸
Signature of Employer

____
Print Name of Employer

____
Address of Employer

____
Employer Licensce No.

RETURN TO APPLICANT

WHEN FILED WITH AN APPLICATION, THIS CERTIFICATE BECOMES THE PROPERTY OF THE PCCLB AND IS KEPT AS A MATTER OF RECORD.

7887 Bryan Dairy Road, Suite 133 • Largo, Florida 33777 • Phone 727-582-3100 • www.pcclb.com

Pinellas County Construction Licensing Board

❶ **DATES OF EMPLOYMENT**

❷ **STATEMENT OF APPLICANT'S EXPERIENCE**

❸ **SIGNATURE OF EMPLOYER**

❹ **NOTARIZATION**

Figure 5. *(continued on next page)*

PINELLAS COUNTY
CONSTRUCTION LICENSING BOARD

TO PERSONS REQUESTED TO CERTIFY TO APPLICANT'S EXPERIENCE:

The applicant named is required to prove his right to take a journeyman's examination by furnishing these certificates in support of his/her experience shown in the application. Enough verification is required to prove such experience to the Board. The applicant must have had:

Not less than four (4) years' experience preceding the filing of an application in the particular classification of license for which application is made:	• ELECTRICIANS • PLUMBERS • PIPE FITTERS • AIR CONDITIONING
Not less than three (3) years' experience preceding the filing of an application in the particular classification of license for which application is made:	• SHEET METAL

❺ You are being requested to certify as to your knowledge of applicant experience by completing the attached form. Verifications must be by **licensed contractors** with direct knowledge of the applicant's experience.

THE FORM MUST BE NOTARIZED.

DO NOT MAIL THE FORM TO THE PINELLAS COUNTY CONSTRUCTION LICENSING BOARD. RETURN IT TO THE APPLICANT IN ORDER FOR HIM/HER TO ATTACH TO HIS/HER APPLICATION.

Your cooperation is earnestly solicited so that the PCCLB can determine whether an applicant has had the experience necessary to become a capable and qualified journeyman.

❻ HOW TO COMPUTE YEARS OF EXPERIENCE – USE ANY OF THE FOLLOWING:

1. Completion of a 4-year apprenticeship program equals 4 years; or
2. Completion of 4 years' full-time employment (2,000 hours/year) since leaving (high) school equals 4 years; or
3. Completion of less than 4 years' full-time employment since leaving (high) school can be added to actual part-time hours worked before leaving school once age 16 was attained. This means actual part-time hours worked since a 16th birthday and during school years may be added to years of work since leaving school to total 4 years.
   The total hours needed to qualify for 4 years equals 8,000 (4 x 2,000). For the sheet metal journeyman classification only verification of 6,000 hours is required.
   ❼ All work must be verified by licensed contractors or building officials familiar with the applicant's experience.

7887 Bryan Dairy Road, Suite 133 • Largo, Florida 33777 • Phone 727-582-3100 • www.pcclb.com

Pinellas County Construction Licensing Board

❺ **CERTIFICATION**

❻ **YEARS OF EXPERIENCE**

❼ **VERIFICATION**

Figure 5. The certificate of journeyman applicant's experience qualifications is completed by the employer.

Exam Grading

Each licensing board sets the grading standard for exams. Contact a local licensing board to determine if they provide an exam breakdown and how the exam is graded.

Journeyman electrician's exams may be developed locally by the authority having jurisdiction (AHJ) or by an independent testing agency. The local licensing board sets standards and approves exam questions. They also collect all fees and process all applications.

Exam Specifications

The local licensing board provides specifications on the structure of an exam. **See Figure 6.** A journeyman electrician's exam is typically a three hour exam with 80 questions. It is a broad-based exam containing questions on calculations, trade knowledge, and finding answers to questions in the NEC®.

When taking the exam, time is critical. All questions not answered are marked wrong. A time frame of three hours provides 2 minutes and 15 seconds for each question (180 min ÷ 80 questions = 2 min 15 sec per question). A stopwatch or small travel clock is ideal for timing. It is recommended to practice answering questions with a timer to simulate an examination setting.

What to Take to the Exam

1. National Electrical Code® (ANSI/NFPA 70). Generally no notes, examples, or formulas are permitted to be written in the book.
2. Any reference books permitted (e.g., American Electricians' Handbook).
3. A minimum of two lead pencils, or a mechanical pencil with additional lead.
4. A stop watch or small travel clock.
5. A 6″ flat ruler for sighting lines in tables.
6. A pad of paper.
7. A calculator (with extra batteries or a spare calculator).

Grade Notification

After completing the journeyman electrician's exam, generally the applicant receives a letter with the grade score and license. A common passing grade is 70% or higher for the total exam. **See Figure 7.**

Reciprocity

Jurisdictions may have mutual agreements allowing journeymen to reciprocate their exam scores. This permits journeymen to perform work in another jurisdiction without retesting. Reciprocity generally requires a higher test score than the minimum passing grade. A score of 75% or higher is common.

Test Specification Sample

Journeyman Electrician's Exam

Scope

The Journeyman Electrician's Exam tests for competency of the installation, maintenance, modification, planning, and repair of electrical conductors, raceways, and equipment.

Time limit **3 hr**

Questions **80** **70%** needed to pass exam

THIS EXAM MAY INCLUDE ADDITIONAL NONSCORED TRIAL QUESTIONS

SUBJECT	PERCENT
General Electrical Knowledge	10
Wiring and Protection	25
Wiring Methods and Materials	20
Equipment for General Use	20
Special Occupancies	10
Special Equipment	5
Special Conditions	5
Communications Systems	5

The following references are permitted during the exam:

- *NFPA 70®, National Electrical Code®*, 2020 Edition
  National Fire Protection Association®
  1 Batterymarch Park
  P. O. Box 9101
  Quincy, MA 02169-9101, www.nfpa.org

- *Printreading Based on the 2020 National Electrical Code® (NEC®)*, 2020 Edition
  American Technical Publishers, Inc.
  10100 Orland Parkway, Suite 200
  Orland Park, IL 60467, www.atplearning.com

Figure 6. *(continued on next page)*

Sample Information Bulletin

Journeyman Electrician's Exam

License Information

The Journeyman Electrician's Exam tests for competency of the installation, maintenance, modification, planning, and repair of electrical conductors, raceways, and equipment.

Exam Outline

80 Questions
3 hr time limit
70% needed to pass exam.

Topic	%
General Electrical Knowledge	10
Wiring and Protection	25
Wiring Methods and Materials	20
Equipment for General Use	20
Special Occupancies	10
Special Equipment	5
Special Conditions	5
Communications Systems	5

The following references are permitted during the exam:

NFPA 70®, National Electrical Code®, 2020 Edition
National Fire Protection Association®
1 Batterymarch Park
P. O. Box 9101
Quincy, MA 02169-9101, www.nfpa.org

Printreading Based on the 2020 National Electrical Code® (NEC®), 2020 Edition
American Technical Publishers, Inc.
10100 Orland Parkway, Suite 200
Orland Park, IL 60467, www.atplearning.com

Figure 6. *(continued on next page)*

SAMPLE QUESTIONS

1. A 5 HP, 240 V, 1ϕ motor drawing 19 A is ___% efficient.
A. 82
B. 89
C. 92
D. 96
Answer: A

2. The ampacity of a 12 AWG Type NM Cu cable used for derating is ___ A.
A. 15
B. 25
C. 30
D. 16
Answer: C

3. Six feet of track lighting in a dwelling shall be computed at ___ VA.
A. 900
B. 0
C. 1080
D. 600
Answer: B

4. The maximum spacing for receptacles in a dwelling is ___′.
A. 6
B. 12
C. 5
D. 10
Answer: B

5. An underground service is called a service ___.
A. drop
B. point
C. lateral
D. cable
Answer: C

6. A ___ is not permitted to be attached to a service riser.
A. telephone line
B. cable TV line
C. computer service
D. all of the above
Answer: D

Figure 6. Exam specifications provide a list of topics covered on the exam along with the number of questions for each topic.

PINELLAS COUNTY
CONSTRUCTION LICENSING BOARD

Date

Dear ________________:

1 On behalf of the Pinellas County Construction Licensing Board, I am pleased to inform you that you have passed the examination which you recently took for Electrical Journeyman with a grade of 76.0%.

2 Enclosed is your license which must be renewed by October 1 of each year.

Very truly yours,

Executive Director

7887 Bryan Dairy Road, Suite 133 • Largo, Florida 33777 • Phone 727-582-3100 • www.pcclb.com

Pinellas County Construction Licensing Board

1 **GRADE SCORE**

2 **LICENSE**

Figure 7. A common passing grade is 70% or higher for a journeyman electrician's exam.

Learner Resources
atplearningresources.com/quicklinks
Access Code: 533920

CHAPTER 1

NEC® Definitions

Definitions of terms used in two or more articles in the NEC® are found in Article 100. Definitions are also listed in other articles of the NEC® and may be referenced in Article 100. Trade names and construction slang are not listed or used in the NEC®.

DEFINITIONS

Definitions of terms used in more than one article of the NEC® are found in Article 100. This article is constructed in three parts. Part I applies to definitions throughout the entirety of the NEC®. Part II applies to definitions for equipment and installations operating over 1000 V. Part III applies to definitions for equipment and installations in hazardous (classified) locations.

Definitions specific to only one article of the code are included in the applicable code section under section 2 (NEC® XXX.2). For example, a fixed appliance is defined in Section 550.2 and a park trailer is defined in 552.2. Trade names and construction slang are not listed or used in the NEC®. For example, Romex is a trade name of nonmetallic-sheathed cable (NMC). Also, electricians commonly refer to the ungrounded conductor as the hot or phase conductor.

Part I—General

Accessible (as pertains to equipment): Capable of being reached for operation, renewal, and inspection. **See Figure 1-1.**

Example: A disconnect for an A/H in an attic with an access hatch is considered *accessible*.

Ampacity: The maximum current, in amperes, that a conductor can carry continuously under the conditions of use without exceeding its temperature rating.

Example: The *ampacities* of bare or covered conductors in free air are listed in Table 310.21.

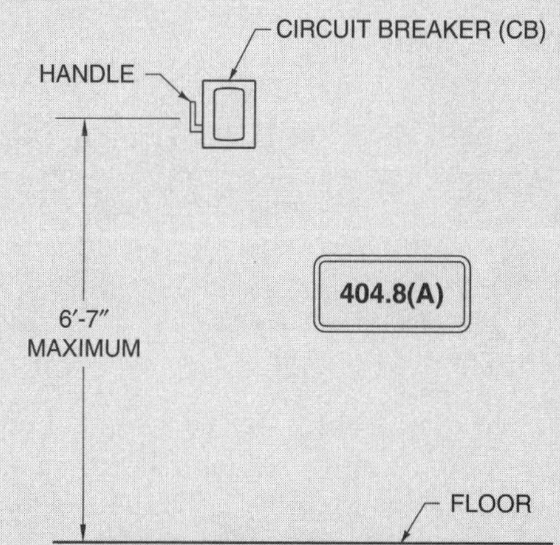

Figure 1-1. Switches and CBs used as switches shall be accessible and not more than 6′-7″ from the floor.

Approved: Acceptable to the authority having jurisdiction (AHJ).

Example: A lay-in fixture surface mounted on a drywall ceiling was not *approved* by the inspector.

Attachment Plug (Plug Cap or Plug): A device that, by insertion in a receptacle, establishes a connection between the conductors of the attached flexible cord and the conductors connected permanently to the receptacle.

Example: An *attachment plug* is NEMA-rated according to voltage and amperage. The two basic plug types are straight blade and twist lock. NEMA charts are available at local supply houses.

Bonded (Bonding): Connected to establish electrical continuity and conductivity. **See Figure 1-2.**

Example: The NEC® requires *bonding* of certain equipment to establish electrical continuity and conductivity. Service equipment with metal raceways shall be bonded. Locknuts alone are not acceptable. Often, bonding bushings are utilized to ensure proper continuity.

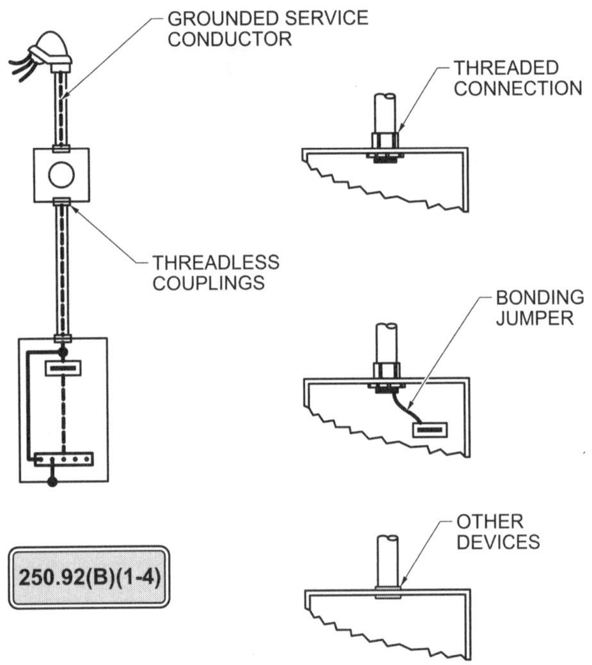

Figure 1-2. Electrical continuity at service equipment is ensured by bonding.

Bonding Jumper, Main: The connection between the grounded circuit conductor and the equipment grounding conductor (EGC), or the supply-side bonding jumper, or both, at the service.

Example: The *main bonding jumper* can be a wire, screw, or an acceptable conductor per 250.102(A).

Branch Circuit: The circuit conductors between the final overcurrent device protecting the circuit and the outlet(s).

Example: Rules and regulations regarding a *branch circuit* are covered in Article 210.

Building: A structure that stands alone or that is separated from adjoining structures by fire walls.

Example: Each house of a block of row houses with approved fire walls is considered a *building*.

Conductor, Bare: A conductor having no covering or electrical insulation whatsoever.

Example: Panelboards often employ a *bare conductor* bus, often referred to as a ground bar, along each section to provide continuous grounding and bonding of the complete panelboard.

Note: "Conductor" does not only denote wire, cable, or bars. A conductor can be the conduit utilized to enclose the circuit conductors. Thus, a conduit run could be an equipment grounding conductor (EGC).

Conductor, Covered: A conductor encased within material of composition or thickness that is not recognized by the NEC® as electrical insulation.

Example: The EGC sleeved in paper within an NM or MC cable would be considered *covered*, not insulated.

Note: "Conductor" does not only denote wire or cable. A conductor can be the components that enclose the live portions of a circuit. Thus, painted panel boxes, liquidtight flexible metal conduit (LFMC), and other components could be considered covered conductors.

Conductor, Insulated: A conductor encased within material of composition and thickness that is recognized by the NEC® as electrical insulation.

Example: The circuit conductors within an NM or MC cable are required to be *insulated* for the appropriate system voltages for which they are utilized.

Note: "Conductor" does not only denote wire or cable. An insulated conductor could be the properly separated (via material, spacing, or both) busbar conductors of a busway system.

Continuous Load: A load where the maximum current is expected to continue for 3 hours or more. **See Figure 1-3.**

Example: Lighting and heating loads are considered *continuous loads* because they may remain active for 3 hours or more.

Device: A unit of an electrical system, other than a conductor, that carries or controls electrical energy as its principal function.

Example: Wall switches, dimmers, duplex receptacles, fuses, and circuit breakers are typical *devices* used in an electrical system.

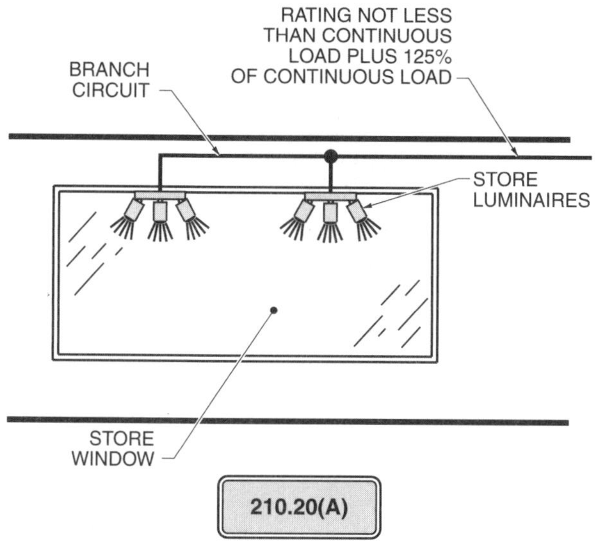

Figure 1-3. Show window luminaires are considered a continuous load.

Duty, Continuous: Operation at a substantially constant load for an indefinitely long time.

Example: Production equipment is often considered *continuous duty* equipment.

Note: There are five distinct definitions of duty that relate to the amount of time a load may be active and to what extent—continuous duty, intermittent duty, periodic duty, short-time duty, and varying duty.

Dwelling Unit: A single unit, providing complete and independent living facilities for one or more persons, including permanent provisions for living, sleeping, cooking, and sanitation.

Example: Long-stay motels or hotels that include cooking areas are considered *dwelling units.*

Note: There are three distinct definitions of dwellings prior to "dwelling unit" that identify specific uses in the NEC®— one-family dwelling, two-family dwelling, and multifamily dwelling.

Electric Sign: A fixed, stationary, or portable self-contained, electrically operated and/or electrically illuminated utilization equipment with words or symbols designed to convey information or attract attention.

Example: Most AHJs require that an *electric sign* be installed by a licensed sign contractor. The electrician provides wiring to the installation.

Energized: Electrically connected to, or is, a source of voltage.

Example: When an electrical device is turned on, it becomes *energized.*

Exposed (as applied to live parts): Capable of being inadvertently touched or approached nearer than a safe distance by a person.

Example: A panel on a construction site used for temporary power has no cover. The wiring and equipment *exposed* presents a hazard.

Exposed (as applied to wiring methods): On or attached to the surface or behind panels designed to allow access.

Example: Junction boxes contained within secured spaces of the equipment proper are considered *exposed* for wiring purposes.

Note: Unlike the term "exposed (as applied to live parts)," where the concern is inadvertent approach, the requirement to remove covers or open panels in the term "exposed (as applied to wiring methods)" denotes that the terminations within junction boxes are considered exposed when at the junction box location.

Feeder: All circuit conductors between the service equipment, the source of a separately derived system, or other power supply source and the final branch-circuit overcurrent device.

Example: Article 215 covers the installation requirements, overcurrent protection requirements, minimum size, and ampacity of conductors for *feeders.*

Fitting: An accessory such as a locknut, bushing, or other part of a wiring system that is intended primarily to perform a mechanical rather than an electrical function.

Example: A conduit connector or coupling is considered a *fitting.*

Note: Conduit bodies (Type T, LB, etc.) are referred to as such because they are not primarily utilized to mechanically connect conduits but rather to provide a point of access to the internal conduit system.

Grounded (Grounding): Connected (connecting) to ground or to a conductive body that extends the ground connection. **See Figure 1-4.**

Example: Electrodes, metal underground pipe, metal in-ground structures, and ground rings are *grounded* in direct contact with the earth per 250.52.

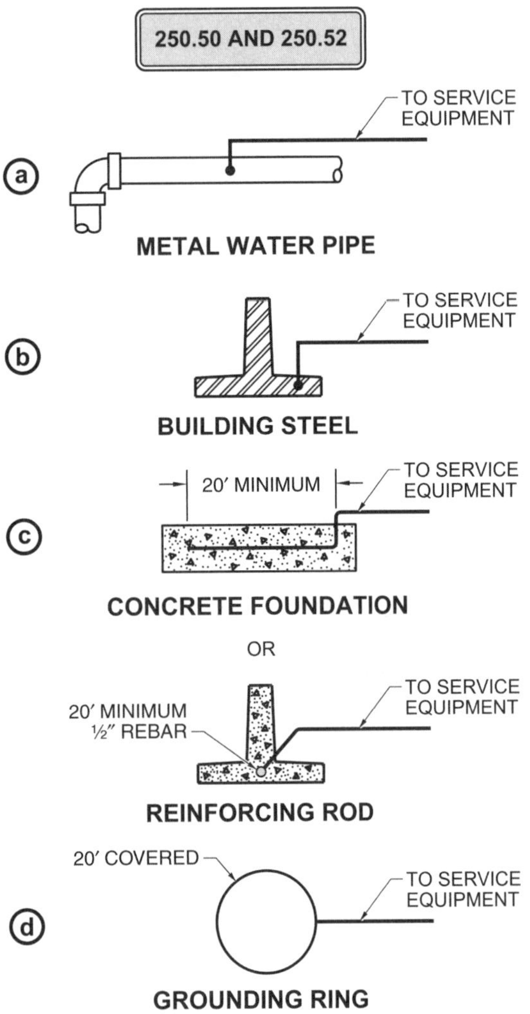

Figure 1-4. Electrical service equipment is grounded per 250.50.

Grounded Conductor: A system or circuit conductor that is intentionally grounded.

Example: A neutral conductor is a *grounded conductor;* the B phase of a grounded B phase system is also a *grounded conductor.*

Note: A grounded conductor is anticipated to carry normal electrical current because it is a part of the working circuit.

Grounding Conductor, Equipment (EGC): A conductive path(s) that is part of an effective ground-fault current path and connects normally non-current-carrying metal parts of equipment together and to the system grounded conductor or to the grounding electrode conductor, or both. **See Figure 1-5.**

Informational Note No. 1: It is recognized that the equipment grounding conductor also performs bonding.

Informational Note No. 2: See 250.118 for a list of acceptable equipment grounding conductors.

Example: The *equipment grounding conductor* of a 120 V, 15 A grounded cord is connected to the round grounding pin of the cord cap.

Note: An equipment grounding conductor, often called the grounding conductor, is not part of the working circuit but is utilized to carry stray currents, often called grounds, back to its source. Its low impedance and intimate contact with conductive materials allows for quick response by overcurrent protective devices. The names "equipment grounding conductor (EGC)," "grounding conductor," and "ground" seem to imply a wire conductor, the NEC® does not mandate that the EGC be a wire. It may be a conduit or raceway that serves as the path back to the source for stray currents.

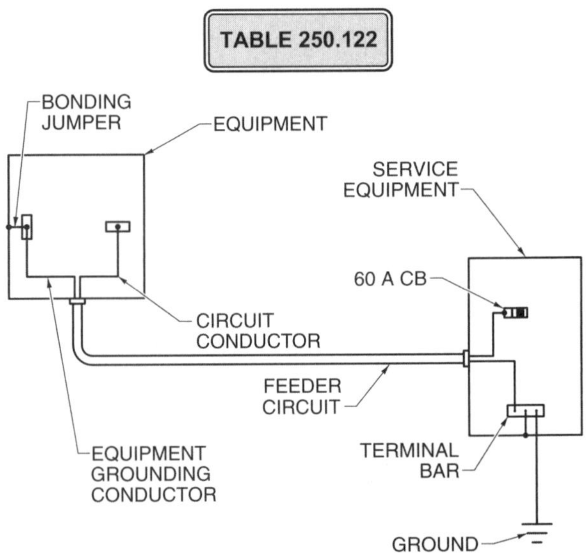

Figure 1-5. An EGC is sized per Table 250.122.

Guarded: Covered, shielded, fenced, enclosed, or otherwise protected by means of suitable covers, casings, barriers, rails, screens, mats, or platforms to remove the likelihood of approach or contact by persons or objects to a point of danger.

Example: A room with a locked door is considered *guarded.*

Handhole Enclosure: An enclosure for use in underground systems, provided with an open or closed bottom, and sized to allow personnel to reach into, but not enter, for the purpose of installing, operating, or maintaining equipment or wiring or both.

Example: Handhole enclosures are common for irrigation systems where all connections are made below grade.

Hoistway: Any shaftway, hatchway, well hole, or other vertical opening or space in which an elevator or dumbwaiter is designed to operate.

Example: Many *hoistways* have a sump in the bottom of the *hoistway* because they are typically below grade and are subject to water intrusion.

In Sight From (Within Sight From, Within Sight): Where the NEC® specifies that one equipment shall be "in sight from," "within sight from," or "within sight of," and so forth, another equipment, the specified equipment is to be visible and not more than 15 m (50′) distant from the other. **See Figure 1-6.**

Example: Many disconnecting devices are required *within sight* of the equipment to ensure greater safety for the technician providing maintenance or repairs.

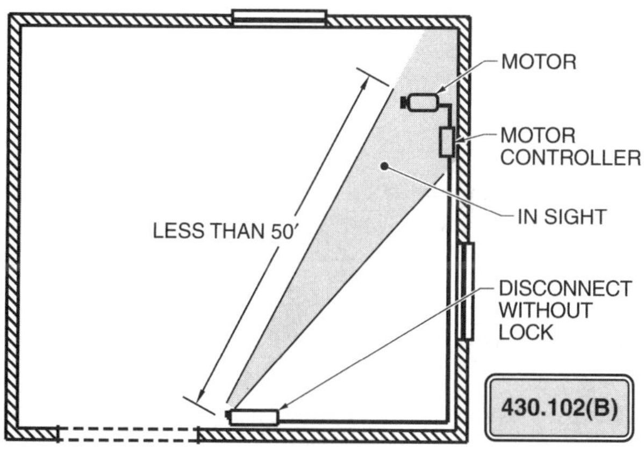

Figure 1-6. If the disconnect cannot be locked in an open position, it shall be in sight of the motor.

Interrupting Rating: The highest current at rated voltage that a device is identified to interrupt under standard test conditions.

Informational Note: Equipment intended to interrupt current at other than fault levels may have its interrupting rating implied in other ratings, such as horsepower or locked rotor current.

Example: Serious damage could occur if an overcurrent protective device (OCPD) is undersized and the fault current exceeds the *interrupting rating*.

Labeled: Equipment or materials to which has been attached a label, symbol, or other identifying mark of an organization that is acceptable to the authority having jurisdiction (AHJ) and concerned with product evaluation, that maintains periodic inspection of production of labeled equipment or materials, and by whose labeling the manufacturer indicates compliance with appropriate standards or performance in a specified manner.

Example: Equipment *labeled* UL meets the standards of Underwriters' Laboratories, Inc.

Live Parts: Energized conductive components.

Example: When work is performed inside a power panel, the bus and any points of connection are considered energized conductive components, or *live parts*.

Nonlinear Load: A load where the wave shape of the steady-state current does not follow the wave shape of the applied voltage.

Informational Note: Electronic equipment, electronic/electric-discharge lighting, adjustable-speed drive systems, and similar equipment may be nonlinear loads.

Example: A *nonlinear load* is a load in which the voltage and current are out-of-phase. Generally, the current is lagging.

Overcurrent: Any current in excess of the rated current of equipment or the ampacity of a conductor. It may result from overload, short circuit, or ground fault.

Informational Note: A current in excess of rating may be accommodated by certain equipment and conductors for a given set of conditions. Therefore, the rules of overcurrent protection are specific for particular situations.

Example: When a motor is bogged down, its *overcurrent* is considered an overload; when two circuit wires are shorted together, the *overcurrent* is considered a short circuit; and when any ungrounded conductor shorts to ground and causes an *overcurrent,* it is considered a ground fault.

Qualified Person: One who has skills and knowledge related to the construction and operation of electrical equipment and installations and has received safety training to recognize and avoid the hazards involved.

Informational Note: Refer to the current edition of NFPA 70E, *Standard for Electrical Safety in the Workplace,* for electrical safety training requirements.

Example: A licensed journeyman electrician is a *qualified person.*

Raintight: Constructed or protected so that exposure to a beating rain will not result in the entrance of water under specified test conditions.

Example: A disconnect installed out of doors in the open must be *raintight.*

Receptacle Outlet: An outlet where one or more receptacles are installed.

Example: Hallways 10′ or more in length shall have at least one *receptacle outlet* per 210.52(H).

Service Drop: The overhead conductors between the serving utility and the service point.

Example: Service drop conductors are installed by private utility companies.

Service Lateral: The underground conductors between the utility electric supply system and the service point.

Example: Most new dwellings are being built with a *service lateral.*

Service Point: The point of connection between the facilities of the serving utility and the premises wiring.

Informational Note: The service point can be described as the point of demarcation between where the serving utility ends, and the premises wiring begins.

Example: The *service point* for the typical overhead service is at the service head and is made via crimp connections by the utility personnel.

Show Window: Any window, including windows above doors, used or designed to be used for the display of goods or advertising material, whether it is fully or partly enclosed or entirely open at the rear and whether or not it has a platform raised higher than the street floor level.

Example: A *show window* that is 12′ or longer requires a receptacle above the window.

Special Permission: The written consent of the authority having jurisdiction (AHJ).

Example: The AHJ may grant special permission per 90.4.

Thermal Protector (as applied to motors): A protective device for assembly as an integral part of a motor or motor-compressor that, when properly applied, protects the motor against dangerous overheating due to overload and failure to start.

Example: A fractional HP motor generally has a built-in *thermal protector.*

Informational Note: The thermal protector may consist of one or more sensing elements integral with the motor or motor-compressor and an external control device.

Voltage, Nominal: A nominal value assigned to a circuit or system for the purpose of conveniently designating its voltage class (e.g. 120/240 V, 480Y/277 V, 600 V).

Informational Note No. 1: The actual voltage at which a circuit operates can vary from the nominal within a range that permits satisfactory operation of equipment.

Informational Note No. 2: See ANSI C84.1-2011, *Voltage Ratings for Electric Power Systems and Equipment (60 Hz).*

Informational Note No. 3: Certain battery units may be considered to be rated at nominal 48 VDC, but may have a charging float voltage up to 58 V. In DC applications, 60 V is used to cover the entire range of float voltages.

Example: The *nominal voltage* for most residential services is considered 120/240 V.

Watertight: Constructed so that moisture will not enter the enclosure under specified test conditions.

Example: A swimming pool light is required to be *watertight.*

Part II—Over 1000 V, Nominal

Multiple Fuse: An assembly of two or more single-pole fuses.

Example: On occasion, two fuses may be physically connected by a fuse manufacturer, producing a *multiple fuse* to allow the proper fusing beyond the largest fuse manufactured.

Substation: An assemblage of equipment (e.g., switches, interrupting devices, circuit breakers, buses, and transformers) through which electric energy is passed for the purpose of distribution, switching, or modifying its characteristics.

Example: Many large industrial complexes utilize outdoor *substations* for ease of distribution between buildings.

Part III—Hazardous (Classified) Locations

Combustible Gas Detection System: A protection technique utilizing stationary gas detectors in industrial establishments.

Example: Gas detectors are used to ensure that atmospheres normally considered safe have not been infiltrated by undetected leaks.

Oil Immersion: Electrical equipment immersed in a protective liquid in such a way that an explosive atmosphere that may be above the liquid or outside the enclosure cannot be ignited.

Example: Contacts of a relay may be *immersed in oil* to preclude electrical sparks in the atmosphere upon make and break conditions.

CHAPTER 1

NEC® Definitions

Review Questions 1-1

Name ______________________ **Date** ______________

The answers for all review questions are listed in the answer key located in the back of this book.

______________ **1.** Automatic is self-acting, functioning by its own mechanism when actuated by some detached influence, such as ___.

A. a change in current strength C. mechanical configuration
B. temperature D. without human intervention

______________ **2.** An enclosure designed either for surface or flush mounting and provided with a frame, in which a swinging door may be hung is a ___.

A. cabinet C. switchboard
B. panelboard D. cutout box

______________ **3.** Copper forms a minimum of ___% of the cross-sectional area of a solid copper-clad aluminum wire.

A. 80 C. 10
B. 125 D. 12.5

______________ **4.** Ampacity is expressed in ___.

A. volts × amps C. volts × watts
B. amperes D. ohms

______________ **5.** Intermittent duty is a mode of service that demands operation for alternate intervals of ___.

A. load and no-load C. load, no load, and rest
B. load and rest D. all of the above

______________ **6.** Accessible (as pertains to equipment) means capable of being reached for ___.

A. operation C. inspection
B. renewal D. all of the above

______________ **7.** Handhole enclosures are used for the purpose of ___ equipment in underground systems.

A. installing C. maintaining
B. operating D. all of the above

______________ **8.** The point of connection between the facilities of the serving utility and the premises wiring is the ___.

A. transformer tie C. system connection tie
B. service tie D. service point

_______________ **9.** A load in which the maximum current is expected to continue for 3 hours or more is a ___ load.

A. peak | C. maximum
B. continuous | D. nominal

_______________ **10.** An LB is a ___.

A. conduit body | C. device
B. fitting | D. service elbow

_______________ **11.** A ___ is a device which serves to govern, in some predetermined manner, the electric power delivered to the machinery to which it is attached.

A. switch | C. contactor
B. relay | D. controller

_______________ **12.** A space in which an elevator or dumbwaiter operates is a ___.

A. shaftway | C. well hole
B. hatchway | D. all of the above

_______________ **13.** ___ lighting is an arrangement of electrical lighting designed to call attention to certain items, such as the shape of a building.

A. Festoon | C. Outline
B. Landscape | D. Flood

_______________ **14.** Conduit installed underground is considered ___.

A. in a wet location | C. in a damp location
B. protected | D. covered

_______________ **15.** Island mode is the operational mode for ___.

A. stand-alone power production equipment | C. interconnected microgrids
B. isolated microgrids | D. all of the above

_______________ **16.** A conductor enclosed with a material not considered insulation by the NEC® is ___.

A. encased | C. sheathed
B. covered | D. coated

_______________ **17.** Nominal voltage is ___ voltage.

A. actual | C. demand
B. peak | D. assigned system

_______________ **18.** An enclosure supplied with a continuous flow of protective gas preventing the ingress of a hazardous atmosphere is considered ___.

A. airtight | C. clean
B. raintight | D. pressurized

_______________ **19.** ___ lighting is a string of outdoor lights suspended between two points.

A. Festoon | C. Carnival
B. Outline | D. Pig tail

_______________ **20.** A(n) ___ is a device with a circuit-opening fusible part that is heated and severed by the passage of overcurrent through it.

A. circuit breaker | C. charge controller
B. fuse | D. ampacity

CHAPTER 1

NEC® Definitions

Review Questions 1-2

Name ______________________________ **Date** ______________

The answers for all review questions are listed in the answer key located in the back of this book.

______ **1.** A(n) ___ switch is for use on branch-circuits and is rated in amperage and voltage.

A. general-use snap	C. isolating
B. motor-circuit	D. general-use

______ **2.** A ___ is a device that is designed to open a circuit automatically on a predetermined overcurrent without damage to itself when properly applied within its rating.

A. thermal cutout	C. circuit breaker
B. thermal protector	D. none of the above

______ **3.** ___ has a sealable or lockable entry that prevents access to live parts without opening the enclosure.

A. A cabinet	C. A cutout box
B. Sealable equipment	D. A safe box

______ **4.** Air ducts are connected to the ___ of an air distribution system.

A. shaftway	C. plenum
B. hoistway	D. air handler

______ **5.** A car wash is considered a ___ location.

A. wet	C. dry
B. damp	D. moist

______ **6.** Equipment that been tested and found suitable for a specified purpose or that meets the appropriate designated standards is ___.

A. labeled	C. listed
B. certified	D. approved

______ **7.** A disconnect requiring an extension ladder for servicing is considered ___.

A. isolated	C. unguarded
B. readily accessible	D. restricted

______ **8.** Equipment installed and accessible behind wooden doors is considered ___.

A. exposed	C. covered
B. concealed	D. isolated

______________ **9.** ___ protection is designed to prevent accidental contact with energized parts.
A. Isolated
B. Enclosed
C. Concealed
D. Restricted

______________ **10.** The ___ is a conductor that connects the system grounded conductor or the equipment to a grounding electrode.
A. grounded conductor
B. equipment grounding conductor
C. neutral bond
D. grounding electrode conductor

______________ **11.** A voltage of ___ V is not considered a nominal voltage.
A. 600
B. 277
C. 208
D. none of the above

______________ **12.** A multioutlet assembly is a type of surface, flush, or freestanding raceway designed to hold ___ assembled on the job site or at the factory.
A. lighting outlets
B. lighting outlets and receptacles
C. conductors and receptacles
D. equipment

______________ **13.** Terms appearing in ___ or more articles are listed in Article 100.
A. one
B. two
C. three
D. four

______________ **14.** A(n) ___ conductor may be wrapped with a paper-type material.
A. bare
B. protected
C. insulated
D. covered

______________ **15.** A structure under construction may be considered ___.
A. a damp location
B. buried
C. a dry location
D. a wet location

______________ **16.** A(n) ___ is an unintentional, electrically conductive connection between an ungrounded conductor of an electrical circuit and the normally non-current-carrying conductors, metallic enclosures, metallic raceways, metallic equipment, or earth.
A. ground
B. grounded conductor
C. ground fault
D. effective ground

______________ **17.** ___ is a general term used for fittings, devices, and luminaires for electrical installations.
A. Supplies
B. Equipment
C. Submittals
D. Supply house stock

______________ **18.** Equipment that controls DC voltage or current and can be used to charge a battery is called a ___.
A. charge controller
B. battery charger
C. Wheatstone bridge
D. rectifier

______________ **19.** A ___ is a load in which the maximum current is expected to continue for 3 hours or more.
A. peak load
B. continuous load
C. maximum load
D. continuous duty load

______________ **20.** Construction designed so that dust will not enter the enclosure is ___ construction.
A. sealed
B. dustproof
C. enclosed for dust and fibers
D. dustight

CHAPTER 1

NEC® Definitions

Review Questions 1-3

Name ______________________________ **Date** ______________

The answers for all review questions are listed in the answer key located in the back of this book.

______________ **1.** ___ is the maximum current that a conductor can carry continuously under the conditions of use without exceeding its temperature rating.

A. Fault current | C. Apparent power
B. Ampacity | D. Nominal current

______________ **2.** ___ include(s) fittings, devices, appliances, and luminaires.

A. Material | C. Equipment
B. Electrical parts | D. Apparatus

______________ **3.** A ___ is a type of surface or flush raceway designed to hold wires.

A. floor duct | C. cellular metal raceway
B. multioutlet assembly | D. busway

______________ **4.** A(n) ___ is a service lateral.

A. equipment mounting system | C. floor duct
B. bus duct | D. none of the above

______________ **5.** A service lateral is a set of ___ conductors.

A. drop | C. underground
B. aerial | D. overhead

______________ **6.** A piece of equipment is in sight from another piece of equipment when not more than ___′ apart.

A. 50 | C. 100
B. 75 | D. 25

______________ **7.** The ___ is the ratio of the maximum demand of the system to the total connected load of the system.

A. nameplate | C. continuous load
B. demand factor | D. connected load

______________ **8.** The NEC® recognizes a ___ as a device.

A. switch | C. pilot light
B. light bulb | D. either A or B

________ 9. ___ equipment is not readily accessible to persons unless special means for access are used.

A. Concealed
B. Covered
C. Isolated
D. Dead front

________ 10. A luminaire consists of a ___.

A. housing
B. ballast
C. lamp
D. all of the above

Questions 11–20 pertain to definitions in other articles of the NEC®.

________ 11. The definition for *armored cable* is given in ___.

A. 430.2
B. 551.2
C. 710.2
D. 320.2

________ 12. The definition for a *converter* is given in ___.

A. 430.2
B. 551.2
C. 710.2
D. 700.1

________ 13. The definition for *mobile home* is given in ___.

A. 550.2
B. 810.4
C. 820.1
D. 710.6

________ 14. The definition for *critical branch* is given in ___.

A. 210.3
B. 517.2
C. 514.1
D. 430.1

________ 15. The definition for *feeder assembly* is given in ___.

A. 550.2
B. 215.1
C. 430.2
D. 770.1

________ 16. The definition for *array* is given in ___.

A. 690.2
B. 520.2
C. 517.3
D. 402.1

________ 17. The definition for *busbar* is given in ___.

A. 336.1
B. 386.2
C. 393.2
D. 394.2

________ 18. The definition for *solar cell* is given in ___.

A. 690.2
B. 517.3
C. 700.1
D. 690.1

________ 19. The definition for *wire* is given in ___.

A. 760.1
B. 550.1
C. 805.2
D. 502.1

________ 20. The definition for *hospital* is given in ___.

A. 502.2
B. 550.2
C. 690.1
D. 517.2

Learner Resources
atplearningresources.com/quicklinks
Access Code: 533920

CHAPTER 2

Electrical Formulas

Electrical formulas are used to solve electrical problems and find unknown electrical quantities. By transposing the basic formulas, all unknowns may be found. Ohm's law and the power formula are the most widely used electrical formulas.

OHM'S LAW

Ohm's law is the relationship between the voltage measured in volts (V), current measured in amperes (A), and resistance measured in ohms (Ω) in an electrical circuit. **See Figure 2-1.** Ohm's law states that current in a circuit is proportional to the voltage and inversely proportional to the resistance. Any value in these relationships is found using Ohm's law, which can be written as follows:

$$I = \frac{E}{R}$$

$$E = R \times I$$

$$R = \frac{E}{I}$$

Current Calculation

Current is the flow rate of electrons in a circuit. Current is measured in amperes (A). Current may be direct or alternating. *Direct current (DC)* is current that flows in one direction. *Alternating current (AC)* is current that reverses its direction of flow at regular intervals. To calculate current, apply the following formula:

$$I = \frac{E}{R}$$

where

I = current (in A)

E = voltage (in V)

R = resistance (in Ω)

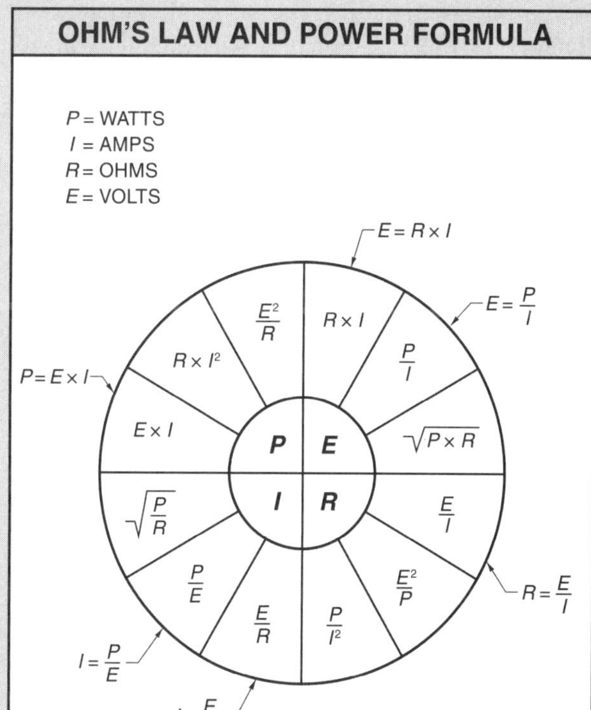

Figure 2-1. The values in the inner circle are equal to values in the corresponding outer circle.

Example: Calculating Current

1. An electrical circuit has a resistance of 80 Ω and is connected to a 120 V supply. Calculate the amount of current in the circuit.

$$I = \frac{E}{R}$$

$$I = \frac{120\text{ V}}{80\ \Omega} = 1.5$$

$$I = \mathbf{1.5\ A}$$

Voltage Calculation

Voltage is the amount of electrical pressure in a circuit. Voltage is measured in volts (V). Voltage is produced by the conversion of chemical energy (battery), light (photocell), electromagnetic energy (generator), heat (thermocouple), pressure (piezo electricity), or friction (static electricity) to electrical energy. Voltage is produced by generating an excess of electrons at one terminal of a voltage source and a deficiency of electrons at the other terminal. The greater the difference in electrons between the terminals, the higher the voltage. To calculate voltage, apply the following formula:

$$E = I \times R$$

where
E = voltage (in V)
I = current (in A)
R = resistance (in Ω)

Example: Calculating Voltage

2. An electrical circuit has 5 A of current and a resistance of 46 Ω. Calculate the voltage in the circuit.

$$E = I \times R$$

$$E = 5\text{ A} \times 46\ \Omega = 230$$

$$E = \mathbf{230\ V}$$

Resistance Calculation

Resistance is the opposition to current flow. Resistance is measured in ohms (Ω). A conductor is material through which current flows easily. Examples of conductors include copper, aluminum, brass, gold, and other metals. An insulator is material through which current cannot flow easily. Examples of insulators include rubber, glass, plastic, paper, varnish, and dry wood. To calculate resistance, apply the following formula:

$$R = \frac{E}{I}$$

where
R = resistance (in Ω)
E = voltage (in V)
I = current (in A)

Example: Calculating Resistance

3. An electrical circuit draws 8 A when connected to a 460 V supply. Calculate the resistance of the circuit.

$$R = \frac{E}{I}$$

$$R = \frac{460\text{ V}}{8\text{ A}} = 57.5$$

$$R = \mathbf{57.5\ \Omega}$$

POWER FORMULA

The power formula is the relationship between the voltage measured in volts (V), current measured in amperes (A), and power measured in volt-amperes (VA) or watts (W) in an electrical circuit. The power formula states that the power in a circuit is equal to the voltage times the current and the power factor (PF). Any value in these relationships is found using the power formula, which can be written as follows:

$$P = E \times I \times PF$$

$$E = \frac{P}{I \times PF}$$

$$I = \frac{P}{E \times PF}$$

The *power factor (PF)* is the electrical efficiency of a circuit determined by the ratio of true power used in an AC circuit to apparent power delivered to the circuit. If all of the current delivered to the load is used to perform the task, the PF is 100% or 1. If only a portion of the current delivered to the load is used to perform the task, then the PF is less than 100% or a decimal value less than 1, with its value being dependent on the actual amount of current used for the task. Therefore, the PF will always be some value between 1 and 0.

Power Calculation

Power is the rate of doing work or using energy. Power may be expressed as true or apparent. *True power* is the actual work being done by the load using electricity. True power (P_T) is measured using a wattmeter and is expressed in watts (W). *Apparent power* is the product of voltage and current in a circuit calculated without considering a phase shift that may be present between total voltage and current in the circuit. Apparent power (P_A) is measured in volt-amperes (VA). A phase shift exists in most AC circuits that contain devices causing capacitance or inductance.

Capacitance is the ability of a component or circuit to store energy in the form of an electrical charge. Inductance is the property of a circuit that causes it to oppose a change in current due to energy stored in a magnetic field. All coils (motor windings, transformers, solenoids, etc.) create inductance in an electrical circuit. True power (P_T) equals apparent power (P_A) in an electrical circuit containing only resistance. True power (P_T) is less than apparent power (P_A) in a circuit containing inductance or capacitance. To calculate apparent power, apply the following formula:

$$P_A = E \times I$$

where

P_A = apparent power (in VA)
E = measured voltage (in V)
I = measured current (in A)

To calculate true power, apply the following formula:

$$P_T = I^2 \times R$$

where

P_T = true power (in W)
I = current (in A)
R = resistance (in Ω)

Example: Calculating Apparent Power

4. An electrical circuit has a measured current of 25 A and a measured voltage of 440 V. Calculate the apparent power of the circuit.

$P_A = E \times I$

$P_A = 440\text{ V} \times 25\text{ A} = 11{,}000$

P_A = **11,000 VA**

Example: Calculating True Power

5. An electrical circuit has a current of 25 A and a total resistive component of 14.75 Ω. Calculate the true power of the circuit.

$P_T = I^2 \times R$

$P_T = 25\text{ A} \times 25\text{ A} \times 14.75\ \Omega$

$P_T = 625 \times 14.75 = 9218.75$

P_T = **9219 W**

Power factor (PF) is the ratio of true power (P_T) used in an AC circuit with respect to the amount of apparent power (P_A) delivered to the circuit. Power factor is expressed as a percentage. True power equals (P_T) apparent power (P_A) when the power factor is 100%. When the power factor is less than 100%, the circuit is less efficient and has higher operating costs. To calculate power factor, apply the following formula:

$$PF = \frac{P_T}{P_A}$$

where

PF = power factor (in %)
P_T = true power (in W)
P_A = apparent power (in VA)

Examples: Calculating Power Factor

6. The apparent power of an electrical circuit is 11,000 VA and the true power is 9219 W. Calculate the power factor of the circuit.

$$PF = \frac{P_T}{P_A}$$

$$PF = \frac{9219\text{ W}}{11{,}000\text{ VA}} \times 100 = 83.8$$

PF = **83.8%**

7. A 120 V appliance is rated at 1200 W. What is the current?

$$I = \frac{P}{E}$$

$$I = \frac{1200\text{ W}}{120\text{ V}} = 10$$

I = **10 A**

8. A 120 V appliance is rated at 1200 W. What is the resistance?

$$R = \frac{E^2}{P}$$

$$R = \frac{120\text{ V} \times 120\text{ V}}{1200\text{ W}} = 12$$

R = **12 Ω**

9. An appliance draws 10 A and has a resistance of 12 Ω. What is the wattage?

$P = I^2 \times R$

$P = 10\text{ A} \times 10\text{ A} \times 12\ \Omega = 1200$

P = **1200 W**

10. A 120 V appliance has 12 Ω of resistance. What is the wattage?

$$P = \frac{E^2}{R}$$

$$P = \frac{120\ \text{V} \times 120\ \text{V}}{12\ \Omega} = 1200$$

$P = \mathbf{1200\ W}$

11. An appliance is rated at 1200 W and has a current of 10 A. What is the resistance?

$$R = \frac{P}{I^2}$$

$$R = \frac{1200\ \text{W}}{10\ \text{A} \times 10\ \text{A}} = 12$$

$R = \mathbf{12\ \Omega}$

SERIES CIRCUIT

A *series circuit* is a circuit with only one path for current to flow. The flow must be continuous. In a series-connected circuit with three light bulbs, an open or blown light bulb causes all lamps to cease operating. **See Figure 2-2.**

SERIES CIRCUIT – TOTAL RESISTANCE

A series circuit has resistances of 500 Ω, 700 Ω, and 100 Ω. Calculate the total resistance in the circuit.

$R_T = R_1 + R_2 + R_3$

$R_T = 500\ \Omega + 700\ \Omega + 100\ \Omega = 1300$

$R_T = \mathbf{1300\ \Omega}$

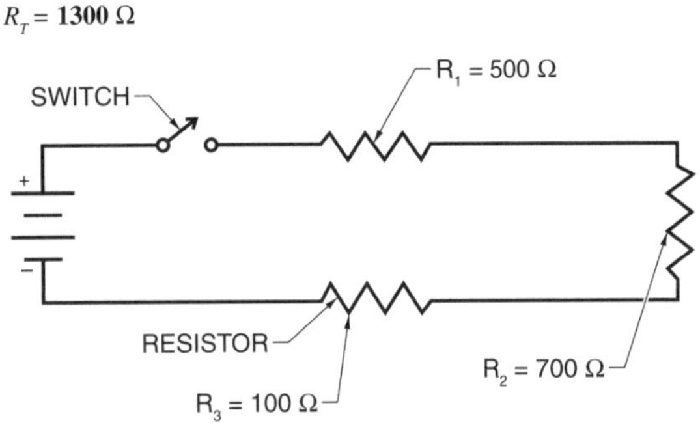

Figure 2-2. A series circuit has only one path for current to flow.

E_T – In a series circuit, the sum of all the voltage drops equal the total voltage applied.

$E_T = E_1 + E_2 + E_3$

I_T – In a series circuit, the current is the same in all parts of the circuit.

$I_T = I_1 = I_2 = I_3$

R_T – In a series circuit, the sum of all the resistances equal the total resistance.

$R_T = R_1 + R_2 + R_3$

Examples: Series Circuit

A series circuit contains a 5 Ω (R_1), 10 Ω (R_2), and 15 Ω (R_3) resistor connected to a 120 V supply.

12. What is the total resistance?

$R_T = R_1 + R_2 + R_3$

$R_T = 5\ \Omega + 10\ \Omega + 15\ \Omega = 30$

$R_T = \mathbf{30\ \Omega}$

13. What is the total current?

$$I_T = \frac{E_T}{R_T}$$

$$I_T = \frac{120\ \text{V}}{30\ \Omega} = 4$$

$I_T = \mathbf{4\ A}$

14. What is the voltage drop?

$E_1 = I_1 \times R_1$

$E_1 = 4\ \text{A} \times 5\ \Omega = 20$

$E_1 = \mathbf{20\ V}$

$E_2 = I_2 \times R_2$

$E_2 = 4\ \text{A} \times 10\ \Omega = 40$

$E_2 = \mathbf{40\ V}$

$E_3 = I_3 \times R_3$

$E_3 = 4\ \text{A} \times 15\ \Omega = 60$

$E_3 = \mathbf{60\ V}$

$E_T = E_1 + E_2 + E_3$

$E_T = 20\ \text{V} + 40\ \text{V} + 60\ \text{V} = 120$

$E_T = \mathbf{120\ V}$

PARALLEL CIRCUITS

The majority of circuits are wired in parallel. Unlike the series-connected circuit, when three light bulbs are connected in parallel and one light bulb blows or opens, the other two will still operate. A *parallel circuit* is a circuit with two or more paths for current to flow. **See Figure 2-3.**

E_T – Voltage is equal in all branches.

$E_T = E_1 = E_2 = E_3$

I_T – In a parallel circuit, the amperage divides according to each parallel resistance. The total current is the sum of all the amperages.

$$I_T = I_1 + I_2 + I_3$$

PARALLEL CIRCUIT – TOTAL RESISTANCE

A parallel circuit has resistances of 2 Ω and 4 Ω. Calculate the total resistance in the circuit.

$$R_T = \frac{R_1 \times R_2}{R_1 + R_2}$$

$$R_T = \frac{2\,\Omega \times 4\,\Omega}{2\,\Omega + 4\,\Omega} = \frac{8}{6} = 1.33$$

R_T = **1.33 Ω**

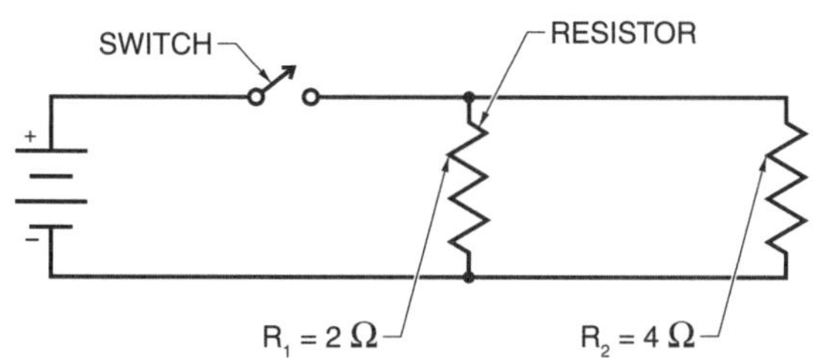

Figure 2-3. A parallel circuit has two or more paths for current to flow.

R_T – In a parallel circuit, the reciprocal of the total resistance equals the sum of the reciprocals of the resistances in each branch. As the resistance of the parallel branches increase, the total resistance decreases. The total resistance in a parallel branch is always less than the smallest resistance. A *reciprocal* is the inverse relationship of two numbers. For example, to find the reciprocal of 6, divide 1 by 6. The reciprocal of 6 is ⅙.

To calculate equal resistors connected in parallel, apply the following formula:

$$R_T = \frac{R_1}{N}$$

N = Number of resistors

To calculate two resistors connected in parallel, apply the following formula:

$$R_T = \frac{R_1 \times R_2}{R_1 + R_2}$$

To calculate three or more resistors connected in parallel, apply the following formula:

$$R_T = \frac{1}{\frac{1}{R_1} + \frac{1}{R_2} + \frac{1}{R_3}}$$

Examples: Parallel Circuits

A circuit contains a 6 Ω (R_1), 8 Ω (R_2), and 12 Ω (R_3) resistor connected in parallel across a 120 V power source.

15. What is E_T?

$$E_T = E_1 = E_2 = E_3$$

$$E_T = \mathbf{120\ V}$$

16. What is R_T?

$$R_T = \frac{1}{\frac{1}{R_1} + \frac{1}{R_2} + \frac{1}{R_3}}$$

$$R_T = \frac{1}{\frac{1}{6\,\Omega} + \frac{1}{8\,\Omega} + \frac{1}{12\,\Omega}}$$

$$R_T = \frac{1}{0.166 + 0.125 + 0.083}$$

$$R_T = \frac{1}{0.374} = 2.67$$

$$R_T = \mathbf{2.67\ \Omega}$$

17. What is I_T?

$$I_T = \frac{E_T}{R_T}$$

$$I_T = \frac{120\text{ V}}{2.67\,\Omega} = 44.9$$

$$I_T = \mathbf{45\ A}$$

18. What is I_1, I_2, and I_3?

$$I_1 = \frac{E_1}{R_1}$$

$$I_1 = \frac{120\text{ V}}{6\,\Omega} = 20$$

$$I_1 = \mathbf{20\ A}$$

$$I_2 = \frac{E_2}{R_2}$$

$$I_2 = \frac{120\text{ V}}{8\,\Omega} = 15$$

$$I_2 = \mathbf{15\ A}$$

$$I_3 = \frac{E_3}{R_3}$$

$$I_3 = \frac{120\text{ V}}{12\ \Omega} = 10$$

$$I_3 = \mathbf{10\ A}$$

SERIES/PARALLEL CIRCUITS

Series/parallel circuits are combinations of series and parallel circuits. To solvc for total resistance, the formulas for series and parallel circuits must be applied. The parallel branches must be reduced to an equivalent resistance (R_Q), which is the total resistance for a parallel branch. Reducing a parallel branch to an equivalent resistance value does not change the total resistance of the circuit.

Examples: Series/Parallel Circuits

19. A 5 Ω resistor is connected in series with 6 Ω and 10 Ω resistors connected in parallel. What is the total resistance of the circuit?

1. Solve for parallel branch equivalent resistance.

$$R_Q = \frac{R_1 \times R_2}{R_1 + R_2}$$

$$R_Q = \frac{6\ \Omega \times 10\ \Omega}{6\ \Omega + 10\ \Omega}$$

$$R_Q = \frac{60}{16}$$

$$R_Q = 3.75\ \Omega$$

2. Substitute R_Q for the parallel branch and solve as a series circuit.

$$R_T = R_1 + R_Q$$
$$R_T = 5\ \Omega + 3.75\ \Omega$$
$$R_T = \mathbf{8.75\ \Omega}$$

20. A circuit has a 4 Ω resistor connected in series with four 4 Ω resistors connected in parallel. What is the total circuit resistance?

1. Solve for parallel branch equivalent resistance.

$$R_Q = \frac{R_1}{N}$$

$$R_Q = \frac{4\ \Omega}{4}$$

$$R_Q = \mathbf{1\ \Omega}$$

2. Substitute R_Q for the parallel branch and solve as a series circuit.

$$R_T = R_1 + R_Q$$
$$R_T = 4\ \Omega + 1\ \Omega$$
$$R_T = \mathbf{5\ \Omega}$$

21. A circuit has a 2 Ω resistor connected in series with a 4 Ω, 6 Ω, and an 8 Ω resistor connected in parallel. What is the total circuit resistance?

1. Solve for parallel branch equivalent resistance.

$$R_Q = \frac{1}{\frac{1}{R_1} + \frac{1}{R_2} + \frac{1}{R_3}}$$

$$R_Q = \frac{1}{\frac{1}{4\ \Omega} + \frac{1}{6\ \Omega} + \frac{1}{8\ \Omega}}$$

$$R_Q = \frac{1}{\frac{1}{0.25} + \frac{1}{0.167} + \frac{1}{0.125}}$$

$$R_Q = \frac{1}{0.542}$$

$$R_Q = 1.84\ \Omega$$

2. Substitute R_Q for the parallel branch and solve as a series circuit.

$$R_T = R_1 + R_Q$$
$$R_T = 2\ \Omega + 1.84\ \Omega$$
$$R_T = \mathbf{3.84\ \Omega}$$

HORSEPOWER

Electrical power is the calculation of the amount of work for a set time period. One electrical horsepower equals 746 W. The ft-lb is a measurement of mechanical power per second or minute. The horsepower is also another method to measure mechanical power. One HP equals 33,000 ft-lb per minute. This is the basis for the relationship between mechanical and electrical energy. **See Figure 2-4.**

Electrical horsepower is found by applying the following formula:

$$HP = \frac{I \times E \times E_{ff}}{746\text{ W}}$$

where
HP = horsepower

I = current (in A)
E = voltage (in V)
E_{ff} = efficiency (in %)
746 W = constant

Any value in a formula may be found by transposing the values. For example, to find the FLC of a motor (which is measured in A), apply the following formula:

$$I = \frac{HP \times 746\ W}{E \times E_{ff}}$$

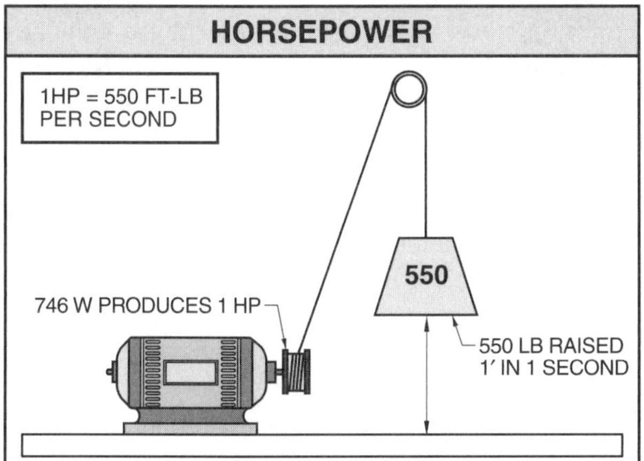

MECHANICAL ENERGY	ELECTRICAL ENERGY
½ HP	373 W
1 HP	746 W
2 HP	1492 W
5 HP	3730 W
100 HP	74,600 W

Figure 2-4. One mechanical horsepower equals 33,000 ft-lb per minute (550 ft-lb per second). One electrical horsepower equals 746 W.

Examples: Horsepower

22. A 20 A, 240 V, single-phase (1ϕ) motor has an 85% efficiency rating. What is the HP?

$$HP = \frac{I \times E \times E_{ff}}{746\ W}$$

$$HP = \frac{20\ A \times 240\ V \times 0.85}{746\ W} = 5.5$$

$HP = \mathbf{5.5\ HP}$

23. A 12 A, 240 V, 1ϕ motor has a 77% efficiency rating. What is the HP?

$$HP = \frac{I \times E \times E_{ff}}{746\ W}$$

$$HP = \frac{12\ A \times 240\ V \times 0.77}{746\ W} = 2.9$$

$HP = \mathbf{2.9\ HP}$

24. What is the FLC of a 5 HP, 230 V motor with a 73% efficiency rating?

$$I = \frac{HP \times 746\ W}{E \times E_{ff}}$$

$$I = \frac{5\ HP \times 746\ W}{230\ V \times 0.73} = 22$$

$I = \mathbf{22\ A}$

25. What is the FLC of a 3 HP, 230 V motor with a 60% efficiency rating?

$$I = \frac{HP \times 746\ W}{E \times E_{ff}}$$

$$I = \frac{3\ HP \times 746\ W}{230\ V \times 0.60} = 16$$

$I = \mathbf{16\ A}$

EFFICIENCY (1Φ MOTORS)

All electrical equipment has losses from heat and friction. *Loss* is the difference between a motor's input and output. For example, a motor with a 1840 VA input and a 746 W output has losses of 1094 VA. *Efficiency* is the output divided by the input of a motor. For example, a 16 A, 1 HP, 115 V, 1ϕ motor may have an efficiency rating of 40%. Efficiency is found by applying the following formula:

$$E_{ff} = \frac{HP \times 746\ W}{I \times E}$$

where

E_{ff} = efficiency (in %)
HP = horsepower
746 W = constant
I = current (in A)
E = voltage (in V)

Efficiency is also found by applying the following formula:

$$E_{ff} = \frac{Output}{Input}$$

where
E_{ff} = efficiency (in %)
$Output = HP \times 746$ (in W)
$Input = V \times A$ (in VA)
Variations of this formula are:

$$Input = \frac{Output}{E_{ff}}$$

$$Output = Input \times E_{ff}$$

Examples: Efficiency

26. What is the efficiency of a 13.2 A, 2 HP, 208 V, 1ϕ motor?

$$E_{ff} = \frac{HP \times 746\text{ W}}{I \times E}$$

$$E_{ff} = \frac{2\text{ HP} \times 746\text{ W}}{13.2\text{ A} \times 208\text{ V}} = 0.54$$

$$E_{ff} = \mathbf{54\%}$$

27. A 2 HP, 240 V, 1ϕ motor draws 8 A. What is the efficiency?

$$E_{ff} = \frac{Output}{Input}$$

$$E_{ff} = \frac{HP \times 746\text{ W}}{V \times A}$$

$$E_{ff} = \frac{2\text{ HP} \times 746\text{ W}}{240\text{ V} \times 8\text{ A}}$$

$$E_{ff} = \frac{1492}{1920} = 0.777$$

$$E_{ff} = \mathbf{77.7\%}$$

28. A 2 HP, 240 V, 1ϕ motor draws 8 A and is 77.7% efficient. What is the input?

$$Input = \frac{Output}{E_{ff}}$$

$$Input = \frac{1492\text{ W}}{0.777} = 1920$$

$$Input = \mathbf{1920\ VA}$$

29. A 2 HP, 240 V, 1ϕ motor draws 8 A and is 77.7% efficient. What is the output?

$Output = Input \times E_{ff}$
$Output = 1920\text{ VA} \times 0.777 = 1492$
$Output = \mathbf{1492\ W}$

30. A 2 HP, 240 V, 1ϕ motor draws 8 A. What are the losses?

$Losses = Input - Output$
$Losses = 1920\text{ VA} - 1492\text{ W} = 428$
$Losses = \mathbf{428\ VA}$

TRANSFORMERS (1Φ)

Transformers operate on the principle of mutual induction. *Mutual induction* is voltage caused in one circuit by a change in current of another circuit. Transformers have a primary winding and a secondary winding. Transformers transfer energy from the primary winding to the secondary winding. Transformers can change the voltage and current, but there is no change of power. The primary winding's power input equals the secondary winding's power output when the transformer is 100% efficient. Transformers are generally considered 100% efficient. Transformers are rated in kVA (kilovolt amps). The k in kVA stands for 1000. When working transformer problems, kVA is often expressed as VA. For example, a 5 kVA transformer is a 5000 VA transformer.

Transformers are very efficient. They are considered 98.2% efficient. In general, for calculation purposes, transformers are considered 100%. In a 100% efficient transformer, the primary kVA equals the secondary kVA, and losses are disregarded.

Transformers of 1000 V or less shall have overcurrent protection per Table 450.3(B). A transformer having a primary current of 9 A or more with an OCPD on the primary rated or set at not more than 125%, does not require an OCPD for the secondary. Where the 125% current rating does correspond to a standard size device, the next higher standard rating shall be permitted per Table 450.3(B), Note 1. See Table 240.6(A).

The current of a transformer's primary is found by applying the following formula:

$$I_P = \frac{kVA \times 1000}{E_P}$$

where
I_P = current of primary (in A)
kVA = kilovolt amps
E_P = voltage of primary (in V)

The current of a transformer's secondary is found by applying the following formula:

$$I_S = \frac{kVA \times 1000}{E_S}$$

where
I_s = current of secondary (in A)
kVA = kilovolt amps
E_s = voltage of secondary (in V)

The kVA of a transformer is found by applying either of the following formulas:

$$kVA = \frac{I_P \times E_P}{1000} \quad \text{or} \quad kVA = \frac{I_S \times E_S}{1000}$$

Examples: Transformers

31. A 5 kVA, 1ϕ, step-up transformer transforms 120 V to 240 V. What is the primary and secondary current? (Efficiency is 100%.)

Primary

$$I_P = \frac{kVA \times 1000}{E_P}$$

$$I_P = \frac{5\text{ kVA} \times 1000}{120\text{ V}}$$

$$I_P = \frac{5000}{120} = 41.6$$

$$I_P = \mathbf{41.6\ A}$$

Secondary

$$I_S = \frac{kVA \times 1000}{E_S}$$

$$I_S = \frac{5\text{ kVA} \times 1000}{240\text{ V}}$$

$$I_S = \frac{5000}{240} = 20.8$$

$$I_S = \mathbf{20.8\ A}$$

32. The secondary voltage of a 1ϕ transformer is 240 V. The current is 12.5 A. What is the primary kVA? (Efficiency is 100%.)

$$kVA = \frac{I_S \times E_S}{1000}$$

$$kVA = \frac{12.5\text{ A} \times 240\text{ V}}{1000}$$

$$kVA = \frac{3000}{1000} = 3$$

$$kVA = \mathbf{3\ kVA}$$

33. A 10 kVA, 1ϕ step-down transformer (240 V to 120 V) is needed to supply a panelboard. What size circuit breaker (CB) is required for primary protection? (Efficiency is 100%.)

$$I_P = \frac{kVA \times 1000}{E_P}$$

$$I_P = \frac{10\text{ kVA} \times 1000}{240\text{ V}}$$

$$I_P = \frac{10{,}000}{240} = 41.6$$

$$I_P = \mathbf{42\ A}$$

Table 450.3(B), Note. 1: 42 A × 125% = 52.5 A
Table 240.6(A): *CB* = **60 A**

34. A 10 kVA, 1ϕ step-down transformer (240 V to 120 V) is needed to supply a panelboard. The secondary overcurrent protection is set at 125% of the secondary FLC. What size CB is required for primary protection? (Efficiency is 100%.)
Table 450.3(B), Note. 1: 42 A × 250% = 105 A
Table 240.6(A): *CB* = **110 A**

Transformer Efficiency

Transformers are very efficient in the range of 85% to 99%. The efficiency increases with the kVA rating of the transformer. Generally, transformers are calculated at 100%. Copper, hysteresis, and eddy currents are types of losses in transformers. To compensate for copper loss, the size of the conductors are increased.

Hysteresis is a lagging in values resulting in a changing magnetization in a magnetic material. To compensate for hysteresis, transformer manufacturers have replaced iron with laminated, silicon steel as the core material.

An *eddy current* is an unwanted, induced current in the core of a transformer. To compensate for eddy currents, manufacturers insulate the laminated sections of the transformer core.

Examples: Transformer Efficiency

35. A 2 kVA, 120/240 V, 1ϕ transformer is 93% efficient. What is the primary power?

$$Input = \frac{Output}{E_{ff}}$$

$$Input = \frac{2000\text{ VA}}{0.93} = 2150$$

$Input =$ **2150 VA**

36. What is the primary and secondary current?

Primary

$$I_P = \frac{P}{E}$$

$$I_P = \frac{2150\text{ VA}}{120\text{ V}} = 17.9$$

$I_P =$ **17.9 A**

Secondary

$$I_P = \frac{P}{E}$$

$$I_P = \frac{2000\text{ VA}}{240\text{ V}} = 8.33$$

$I_P =$ **8.33 A**

37. How much power is lost due to transformer losses?

Losses = *Input* – *Output*

Losses = 2150 VA – 2000 VA = 150

Losses = **150 VA**

Ratio

A *ratio* is the whole numbers designating the relationship between high voltage and low voltage. For example, a transformer with a 120 V primary and a 480 V secondary has a ratio of 1 to 4 (expressed as 1:4).

$$R = \frac{P}{S}$$

where

R = ratio

P = primary voltage

S = secondary voltage

Example: Ratio

38. A transformer has a 120 V primary and a 12 V secondary. What is the ratio?

$$R = \frac{P}{S}$$

$$R = \frac{120}{12} = 10$$

$R =$ **10 : 1**

TEMPERATURE CONVERSION

Temperature is a measurement of the intensity of heat. *Ambient temperature* is the temperature of air around a conductor or a piece of equipment. Current flowing in a conductor produces heat. The heat produced in an electrical circuit may be by design, such as from a heating element, or unintentional, as from a bad conductor splice. Unintentional heat deteriorates insulation and lubrication, and can destroy electrical devices.

Temperature rise is the amount of heat that an electrical component produces above the ambient temperature. All electrical and electronic devices produce heat and are designed to function correctly within a given temperature rise. Temperature is usually measured in degrees Fahrenheit (°F) or degrees Celsius (°C). **See Figure 2-5.**

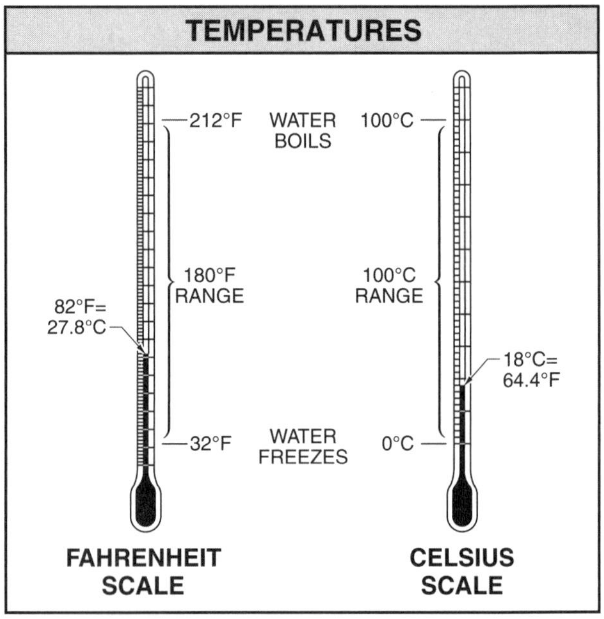

Figure 2-5. Temperature is usually measured in degrees Fahrenheit (°F) or degrees Celsius (°C).

Converting Fahrenheit to Celsius

To convert a Fahrenheit temperature reading to Celsius, subtract 32 from the Fahrenheit reading and divide by 1.8. To convert Fahrenheit to Celsius, apply the following formula:

$$°C = \frac{°F - 32}{1.8}$$

where

$°C$ = degrees Celsius

$°F$ = degrees Fahrenheit

32 = difference between bases

1.8 = ratio between bases

Example: Converting Fahrenheit to Celsius

39. A digital thermometer indicates a reading of 185°F. Convert the Fahrenheit temperature to Celsius.

$$°C = \frac{°F - 32}{1.8}$$

$$°C = \frac{185°\text{F} - 32}{1.8}$$

$$°C = \frac{153}{1.8} = 85$$

$°C =$ **85°C**

Converting Celsius to Fahrenheit

To convert a Celsius temperature reading to Fahrenheit, multiply 1.8 by the Celsius reading and add 32. To convert Celsius to Fahrenheit, apply the following formula:

$°F = (1.8 \times °C) + 32$

where

$°F$ = degrees Fahrenheit

1.8 = ratio between bases

$°C$ = degrees Celsius

32 = difference between bases

Example: Converting Celsius to Fahrenheit

40. A temperature monitor indicates 55°C. Convert the Celsius temperature to Fahrenheit.

$°F = (1.8 \times °C) + 32$

$°F = (1.8 \times 55°\text{C}) + 32$

$°F = 99 + 32 = 131$

$°F =$ **131°F**

BUSBAR CALCULATIONS

Busbars are commonly used in switchboards, load centers, and panelboards. Busbars are generally uninsulated. The constant amperage for a bare copper (Cu) busbar is not permitted to exceed 1000 amperes per square inch (A/sq in.) of the cross sectional area of the bar. For aluminum (Al) busbars, the constant amperage is not permitted to exceed 700 A/sq in. of the cross-sectional area of the bar per 366.23(A). To calculate the ampacity of a busbar, apply the following formula:

busbar ampacity = $t \times w \times$ constant

where

busbar ampacity = ampacity of busbar (in A)

t = thickness (in in.)

w = width (in in.)

constant = 1000 for Cu and 700 for Al

Examples: Calculating Busbar Ampacity

41. What is the ampacity of a ¼″ thick by 3″ wide Cu busbar?

busbar ampacity = $t \times w \times$ constant

busbar ampacity = 0.25″ × 3″ × 1000

busbar ampacity = 0.75 × 1000

busbar ampacity = **750 A**

42. What is the ampacity of a ½″ thick by 2″ wide Al busbar?

busbar ampacity = $t \times w \times$ constant

busbar ampacity = 0.50″ × 2″ × 700

busbar ampacity = 1 × 700

busbar ampacity = **700 A**

VOLTAGE DROP

The NEC® recommends that a voltage drop not exceed 3% of the source voltage for branch circuits per 210.19(A), Informational Note 3. The NEC® recommends that a voltage drop not exceed 3% of the source voltage for feeders per 215.2(A)(1)(b), Informational Note 2. The NEC® recommends that a voltage drop not exceed 5% of the source voltage for a combination of feeders and branch circuits per 215.2(A)(1)(b), Informational Note 2.

The *K factor* is the resistance of a circular mil-foot of wire at a set temperature. K is based on the properties of conductors per Ch 9, Table 8. For example, an uncoated 1000 kcmil Cu conductor at 75°C has a resistance (R) of 0.0129 Ω/kFT per Table 8. To find the resistance (R) for 1′ of this conductor, move the decimal point in 0.0129 (constant/1000′) three places to left. The resistance (R) of 1′ of this conductor is 12.9 μΩ (constant/1′).

Two methods of K factor used in voltage drop calculations are the approximate K factor and the exact K factor. The approximate K factor is most commonly used in voltage drop calculations. Approximate K is a resistance constant of either a copper (Cu) or an aluminum (Al) conductor at 1000 circular mils (CM) that is 1000′ in length and at an operating temperature of 75°C. Approximate K for Cu is 12.9 Ω, whereas the approximate K for Al is 21.2 Ω.

To apply the exact K to the voltage drop formula, the exact K must be known. To find exact K, apply the formula:

$$K = \frac{R \times CM}{1000}$$

where

K = resistance of a conductor at a set temperature (in Ω)

R = resistance of conductors per Ch 9, Table 8 (in Ω/kFT)

CM = circular mils (in area)

1000 = constant

Voltage drop is the voltage that is lost due to the resistance of conductors. The longer the conductor, for a given size, the larger the loss. To compensate for voltage drop, a larger conductor is selected, or the voltage may be increased (if available). Voltage drop may be expressed as a percent. To find the percent of voltage drop, apply the formula:

$$VD = \frac{Line\ Loss}{Supply\ Voltage} \times 100$$

where

VD = voltage drop (in %)

Line Loss = loss of volts

Supply Voltage = source voltage

100 = constant (to obtain %)

Voltage drop is found by applying the formula:

$$VD = \frac{K \times I \times M \times D}{CM}$$

where

VD = voltage drop (in V)

K = resistance of a conductor at a set temperature (in Ω) (Use approximate K.)

I = current (in A)

M = phase multiplier (2 for 1ϕ or 1.73 for 3ϕ)

D = distance (one way) (in feet)

CM = circular mils (in area)

K factor is based on the properties of conductors per Ch 9, Table 8.

Examples: Voltage Drop

43. What is the recommended maximum voltage drop for a 120 V, 208 V, and 240 V branch circuit?

210.19(A), Informational Note 3:

120 V × 3% = **3.6 V**

208 V × 3% = **6.24 V**

240 V × 3% = **7.2 V**

44. What is the exact K for an uncoated 10 AWG solid Cu conductor in a 1ϕ system?

Ch 9, Table 8: uncoated 10 AWG solid Cu R = 1.21 Ω/kFT

Ch 9, Table 8: 10 AWG = 10,380 CM

$$K = \frac{R \times CM}{1000}$$

$$K = \frac{1.21\ \Omega/\text{kFT} \times 10{,}380\ \text{CM}}{1000}$$

$K = \mathbf{12.56\ \Omega}$

45. A 15 A, 120 V, 1ϕ circuit has a wire run of 150′. What size Cu wire is required? (Use approximate K.)

approximate K for Cu = 12.9 Ω

1ϕ multiplier = 2

210.19(A), Informational Note 3: 120 V × 3% = 3.6 V

$$CM = \frac{K \times I \times M \times D}{VD}$$

$$CM = \frac{12.9\ \Omega \times 15\ \text{A} \times 2 \times 150'}{3.6\ \text{V}}$$

$$CM = \frac{58{,}050}{3.6} = 16{,}125$$

Ch 9, Table 8: 8 AWG = 16,510 CM

Cu wire = **8 AWG**

46. A 15 A, 120 V, 1ϕ branch circuit has a wire run of 150′. What is the VD with 10 AWG solid Cu wire? (Use approximate K.)

approximate K for Cu = 12.9 Ω

1ϕ multiplier = 2

Ch 9, Table 8: 10 AWG = 10,380 CM

$$VD = \frac{K \times I \times M \times D}{CM}$$

$$VD = \frac{12.9\ \Omega \times 15\ \text{A} \times 2 \times 150'}{10{,}380\ \text{CM}}$$

$$VD = \frac{58{,}050}{10{,}380} = 5.59$$

$VD = \mathbf{5.59\ V}$

47. A 15 A, 120 V, 1ϕ branch circuit has a wire run of 150′. What is the percentage of VD with 10 AWG solid Cu wire? (Use approximate K.)

$$VD = \frac{Line\ Loss}{Supply\ Voltage} \times 100$$

$$VD = \frac{5.59\ \text{V}}{120\ \text{V}} \times 100 = 4.66$$

$VD = \mathbf{4.66\%}$

48. A 240 V, 1ϕ branch circuit with 8 AWG Cu wire has a 35 A load. What is the length of the two-wire circuit? (Use approximate K and anticipate a maximum recommended voltage drop.)

Ch 9, Table 8: 8 AWG = 16,510 CM

210.19(A), Informational Note 3: 240 V × 3% = 7.2 V

approximate K for Cu = 12.9 Ω

1ϕ multiplier = 2

$$D = \frac{CM \times VD}{K \times I \times 2}$$

$$D = \frac{16{,}510\ \text{CM} \times 7.2\ \text{V}}{12.9\ \Omega \times 35\ \text{A} \times 2}$$

$$D = \frac{118{,}872}{903} = 131.6$$

$D = \mathbf{131.6'}$

49. A 240 V, 1ϕ branch circuit has a 200′ run of 4 AWG Al wire. What is the amperage of the circuit? (Use approximate K.)

Ch 9, Table 8: 4 AWG = 41,740 CM

210.19(A), Informational Note 3: 240 *V* × 3% = 7.2 V

approximate K for Al = 21.2 Ω

1ϕ multiplier = 2

$$I = \frac{VD \times CM}{D \times K \times M}$$

$$I = \frac{7.2\ \text{V} \times 41{,}740\ \text{CM}}{200' \times 21.2\ \Omega \times 2}$$

$$I = \frac{300{,}528}{8480} = 35.4$$

$I = \mathbf{35.4\ A}$

COST OF ENERGY

Electrical energy is provided by the local utility company. The energy used is recorded by a wattmeter. The unit for electrical power is the kilowatt (1000 W). The cost is based on kilowatt-hour (kWh) used. Rates are set by the local utility company.

The cost of energy is found by applying the following formula:

$$Cost = \frac{T \times W \times Cost\,kWh}{1000}$$

where

Cost = price (in dollars and cents)

T = time (in hours)

W = watts

Cost/kWh = price/kWh

1000 = constant

Examples: Cost of Energy

50. What is the cost for operating a 750 W electric heater for 8 hours at 8¢ per kWh?

$$Cost = \frac{T \times W \times Cost\,/\,kWh}{1000}$$

$$Cost = \frac{8\ \text{hr} \times 750\ \text{W} \times 0.08}{1000} = 0.48$$

$Cost = \mathbf{48¢}$

51. What is the cost for operating a 200 W lamp for 24 hours if the utility charge is 13¢ per kWh?

$$Cost = \frac{T \times W \times Cost\,/\,kWh}{1000}$$

$$Cost = \frac{24\ \text{hr} \times 200\ \text{W} \times 0.13}{1000} = 0.62$$

$Cost = \mathbf{62¢}$

TRANSPOSING FORMULAS

The journeyman electrician's exam generally provides base formulas for calculations. To solve for any variable, the base formula is transposed to isolate the variable on one side of the equation. For example, the Ohm's Law formula may be transposed to solve for either voltage, current, or resistance if the other two values are known.

Base formula: $E = I \times R$

where

E = voltage (in V)

I = current (in A)

R = resistance (in Ω)

To solve for current, the base formula is transposed:

Base formula: $E = I \times R$

Divide each side by R (to cancel R from right side):

$$\frac{E}{R} = \frac{I \times R}{R}$$

Final transposed formula: $\frac{E}{R} = I$

or

$$\boldsymbol{I = \frac{E}{R}}$$

Examples: Transposing Formulas

52. What is the formula for finding resistance when using Ohm's law?

Base formula: $E = I \times R$

Divide each side by I (to cancel I from right side):

$$\frac{E}{I} = \frac{I \times R}{I}$$

Final transposed formula: $\frac{E}{I} = R$

or

$$\boldsymbol{R = \frac{E}{I}}$$

53. What is the formula for finding current when using the voltage drop formula?

Base formula: $VD = \frac{K \times I \times M \times D}{CM}$

Multiply each side by CM (to cancel CM from right side): $VD \times CM = (K \times I \times M \times D)$

Divide each side by K (to cancel K from right side):

$$\frac{VD \times CM}{K} = I \times M \times D$$

Divide each side by M (to cancel M from right side):

$$\frac{VD \times CM}{K \times M} = I \times D$$

Divide each side by D (to cancel D from right side):

$$\frac{VD \times CM}{K \times M \times D} = I$$

Final transposed equation: $\frac{VD \times CM}{K \times M \times D} = I$

or

$$I = \boldsymbol{\frac{VD \times CM}{K \times M \times D}}$$

METRICS

The metric system is the most commonly used measurement system in the world. The English system is primarily used in the United States. The metric system is based on the meter (m), liter (l), and gram (g). The English system uses the inch (in. or ″), foot (ft or ′), pint (pt), gallon (gal.), ounce (oz), and pound (lb) as the basic units of measure. Conversion factors are used to change measurements from one system to the other.

English/Metric Conversion Factors

1 inch (in.) = 25.4 millimeters (mm)
1 inch (in.) = 2.54 centimeters (cm)
1 foot (ft) = 0.3048 meters (m)
1 millimeter (mm) = 0.03937 inches (in.)
1 centimeter (cm) = 0.3937 inches (in.)
1 meter (m) = 3.28084 feet (ft)
1 cubic inch (cu in.) = 16.387064 cubic centimeters (cu cm)
1 cubic centimeter (cu cm) = 0.061024 cubic inches (cu in.)
1 square millimeter (sq mm) = 0.001550 square inches (sq in.)
1 square inch (sq in.) = 645.16 square millimeters (sq mm)

Examples: Metrics

54. What is the diameter in inches of a light fixture with a diameter of 17.145 cm?
17.145 cm ÷ 2.54 = **6.75 or 6¾″**

55. What is the length in feet of a length of wire 50 m long?
50 m × 3.28084 = **164′**

56. What is the volume in cubic inches of a box with a volume of 295 cu cm?
295 cu cm × 0.061024 = **18 cu in.**

57. What is the diameter in inches of 10 AWG solid copper wire that has a diameter of 2.588 mm?
2.588 mm × 0.03937 = **0.102″**

58. What is the resistance of 10 m of 1 AWG Cu wire in ohms per meter?
Ch 9, Table 8: resistance of 1 AWG Cu wire = 0.505 Ω/km
0.505 ÷ 1000 = **0.000505 Ω/meter**

59. What is the resistance of 10 m of 1 AWG Cu wire in ohms per foot.
10 m × 3.28084 = 32.8′
Ch 9, Table 8: resistance of 1 AWG Cu wire = 0.154 Ω/kFT
0.154 ÷ 1000 = 0.000154 Ω/ft
0.000154 × 32.8′ = **0.005 Ω/ft**

60. What is the cross-sectional area in millimeters of a ½″ EMT conduit with a cross-sectional area of 0.622″?
0.622″ × 25.4 = **15.8 mm**

61. What is the cross-sectional area in inches of a ¾″ IMC conduit with a cross-sectional area of 377 sq mm?
377 sq mm × 0.001550 = **0.584 sq in.**

CHAPTER 2

Electrical Formulas

Review Questions 2-1

Name ______________________________ **Date** ____________

The answers for all review questions are listed in the answer key located in the back of this book.

________ **1.** ___ is the ratio of true power used in an AC circuit to apparent power delivered to the circuit.

A. Power factor
B. Horsepower
C. Output
D. Input

________ **2.** True power equals apparent power in an electrical circuit containing only ___.

A. resistance
B. capacitance
C. inductance
D. impedance

________ **3.** The NEC® recommends no more than a ___% VD for branch circuits.

A. 2
B. 3
C. 125
D. 5

________ **4.** Water freezes at ___°C and boils at ___°C at sea level.

A. –23; 212
B. 0; 212
C. –32; 100
D. 0; 100

________ **5.** ___ is the same in a parallel branch.

A. Current
B. Resistance
C. Voltage
D. VA

________ **6.** An electrical circuit has a current of 25 A with a total resistance of 21 Ω. The true power of the circuit is ___ W.

A. 46
B. 525
C. 13,125
D. none of the above

________ **7.** The output of a 5 HP motor is ___ VA.

A. 150
B. 1500
C. 370
D. 3730

________ **8.** An A/H has a 10 kW, 240 V, 1ϕ heat strip. The resistance of the heat strip is ___ Ω.

A. 41.6
B. 576
C. 2.88
D. 5.76

________ **9.** The exact K for an uncoated 1 AWG Cu conductor is ___ Ω.

A. 10.8
B. 10.4
C. 12.6
D. 12.88

_______________ **10.** A 4/0 kcmil conductor has a circular mil area of ___.
A. 0.082 C. 211,600
B. 0.575 D. 0.0535

_______________ **11.** The cost of operating a 200 W lamp for 8 hours at 9¢ per kWh is $___.
A. 1.44 C. 0.44
B. 1.84 D. 0.14

_______________ **12.** A circuit has a 4 Ω, 6 Ω, and 8 Ω resistor connected in series. The total resistance is ___ Ω.
A. 1.8 C. 18
B. 0.542 D. none of the above

_______________ **13.** A circuit has a 4 Ω, 6 Ω, and 8 Ω resistor connected in parallel. The total resistance is ___ Ω.
A. 1.845 C. 0.542
B. 18 D. none of the above

_______________ **14.** A 240 V, 1ϕ motor draws 28 A and is 55% efficient. The rating of the motor is ___ HP.
A. 2 C. 3
B. 7½ D. 5

_______________ **15.** The primary maximum overcurrent protection shall not exceed ___ A for a 1.5 kVA, 120/240 V, 1ϕ step-up transformer.
A. 20 C. 10
B. 15 D. none of the above

_______________ **16.** A 40 A, 240 V, 1ϕ branch circuit has a 210′ run of 8 AWG XHHW Al conductor rated at 75°C. The VD is ___ V. (Use approximate K.)
A. 7.2 V C. 15 V
B. 12 V D. none of the above

_______________ **17.** The total resistance in a circuit that has a 3 Ω resistor connected in series with a 4 Ω and a 3 Ω resistor connected in parallel is ___ Ω.
A. 4.7 C. 3.8
B. 13 D. 6.5

_______________ **18.** The total resistance in a circuit that has a 6 Ω resistor connected in series with a 2 Ω, 5 Ω, and 10 Ω resistor connected in parallel is ___ Ω.
A. 23 C. 7.25
B. 8.9 D. 12

_______________ **19.** The ampacity of a ½″ × 4″ Al busbar is ___ A.
A. 1000 C. 700
B. 1400 D. 1250

_______________ **20.** The ampacity of a ¼″ × 4″ Cu busbar is ___ A.
A. 700 C. 1100
B. 1200 D. 1000

CHAPTER 2

Electrical Formulas

Review Questions 2-2

Name ______________________ **Date** ______________

The answers for all review questions are listed in the answer key located in the back of this book.

______ **1.** A 300 W, 130 V incandescent lamp draws ___ A.
A. 3 B. 2.6
C. 2.1 D. 2.3

______ **2.** A circuit has 1.5 A of current through a 40 Ω resistor. The current is ___ A if the resistance is increased to 50 Ω.
A. 1.5 C. 1.2
B. 2 D. 2.6

______ **3.** Three 8 Ω resistors are connected in series. The total resistance is ___ Ω.
A. 24 C. 8
B. 2.66 D. none of the above

______ **4.** Three 8 Ω resistors are connected in parallel. The total resistance is ___ Ω.
A. 2.66 C. 8
B. 24 D. none of the above

______ **5.** A 5 Ω and a 10 Ω resistor are connected in parallel. The total resistance is ___ Ω.
A. 15 C. 50
B. 3.33 D. none of the above

______ **6.** A 10 Ω heater draws 4 A from a power supply. The power supply is ___ V.
A. 120 C. 40
B. 0.4 D. none of the above

______ **7.** The unit for the measurement of work is the ___.
A. Coulomb C. foot-pound
B. Farad D. watt-hour

______ **8.** A conducting material has a certain resistance at 30°C (___°F).
A. 30 C. 86
B. 68 D. 72

______ **9.** The operating temperature of an insulation should not exceed 104°F (___°C).
A. 40 C. 72
B. 60 D. none of the above

__________ **10.** Current is defined as the flow of ___.

A. protons | C. ohms
B. neutrons | D. electrons

____________ **11.** The maximum recommended line loss for a 40 A, 240 V, 1ϕ branch circuit is ___ V.

A. 12 | C. 3.6
B. 7.2 | D. none of the above

____________ **12.** A 115 V, 1ϕ circuit has an 8 V line loss. The voltage drop is ___%.

A. 14.38 | C. 2.64
B. 6.96 | D. none of the above

____________ **13.** A transformer has a 300 VA, 120 V primary and a 295 VA, 12 V secondary. The transformer is ___% efficient.

A. 95 | C. 100
B. 98 | D. 80

____________ **14.** With a source voltage of 120 V and a load of 6500 VA, the voltage across the load reads 115 V. The power consumed by the conductors is ___ VA.

A. 283 | C. 150
B. 300 | D. 75

____________ **15.** The Joule is the measurement of ___.

A. voltage | C. electrons
B. energy | D. resistance

____________ **16.** Electromotive force is ___.

A. potential difference | C. work
B. energy | D. Coulombs per second

____________ **17.** A length of wire 60 m long equals ___′.

A. 152.4 | C. 197
B. 145.4 | D. 210

____________ **18.** A length of conduit 24″ long equals ___ cm.

A. 61 | C. 78
B. 7.31 | D. 55.4

____________ **19.** The total resistance of a circuit that has a 12 Ω resistor connected in series with six 12 Ω resistors connected in parallel is ___ Ω.

A. 18 | C. 14
B. 8.9 | D. 12.4

____________ **20.** The total resistance of a circuit that has an 8 Ω resistor connected in series with 5 Ω, 7 Ω, and 9 Ω resistors connected in parallel is ___ Ω.

A. 10.2 | C. 29
B. 7.1 | D. 12

CHAPTER 2

Electrical Formulas

Review Questions 2-3

Name ______________________________ **Date** ______________

The answers for all review questions are listed in the answer key located in the back of this book.

A 1ϕ transformer with a 120 V primary and a 12 V secondary with a 300 W load is used for problems 1 through 5.

______ **1.** The turns ratio of the transformer is ___.

A. 1:10
B. 10:1
C. 12:1
D. 1:2

______ **2.** The amperage of the secondary is ___ A.

A. 2.5
B. 12
C. 10
D. 25

______ **3.** The minimum rating required is ___ kVA.

A. 3
B. 300
C. 0.3
D. none of the above

______ **4.** The primary current is ___ A.

A. 25
B. 10
C. 2.5
D. none of the above

______ **5.** The resistance of the secondary circuit is ___ Ω.

A. 25
B. 0.48
C. 2.5
D. none of the above

______ **6.** ___ reduces copper losses.

A. A silicon steel core
B. Laminating the core
C. Increasing copper size
D. Using more windings

______ **7.** The phase-to-neutral voltage for a 3ϕ, low-voltage, wye-connected transformer is ___V.

A. 120
B. 240
C. 208
D. 480

______ **8.** The high leg voltage for a low-voltage, delta-connected transformer is ___ V.

A. 120
B. 240
C. 208
D. 480

______ **9.** The Coulomb is the measurement of ___.

A. voltage
B. energy
C. electrons
D. resistance

A 3-wire, 120/240 V, 1ϕ branch circuit with two parallel loads is used for problems 10 through 16. L1 to the neutral has a 120 V, 25 W load. L2 to the neutral has a 130 V, 100 W load. The neutral opens in the panel.

______ **10.** The voltage applied to the loads is ___ V.

A. 120
B. 130
C. 240
D. all of the above

______ **11.** The resistance of the 25 W load is ___ Ω.

A. 576
B. 0.208
C. 9.6
D. none of the above

______ **12.** The resistance of the 100 W load is ___ Ω.

A. 144
B. 169
C. 2.4
D. 0.833

______ **13.** The total resistance is ___ Ω.

A. 745
B. 12
C. 732
D. 1.041

______ **14.** The amperage drawn on L1 conductor is ___ A.

A. 0.16
B. 10
C. 0.322
D. none of the above

______ **15.** The voltage drop across the 25 W load is ___ V.

A. 120
B. 240
C. 96
D. 185.5

______ **16.** The voltage drop across the 100 W load is ___ V.

A. 54.4
B. 130
C. 144
D. 120

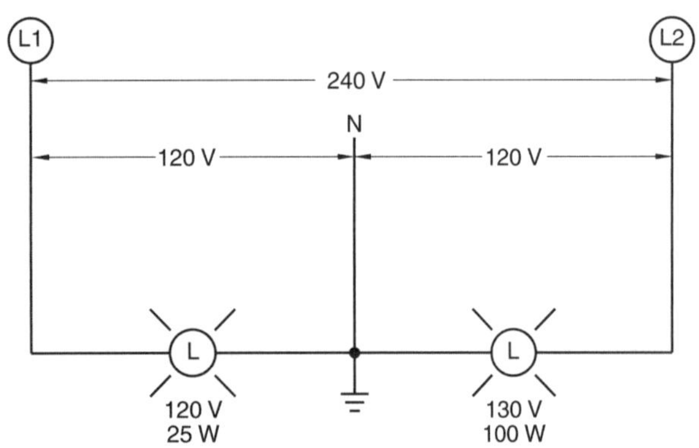

CIRCUIT NORMAL

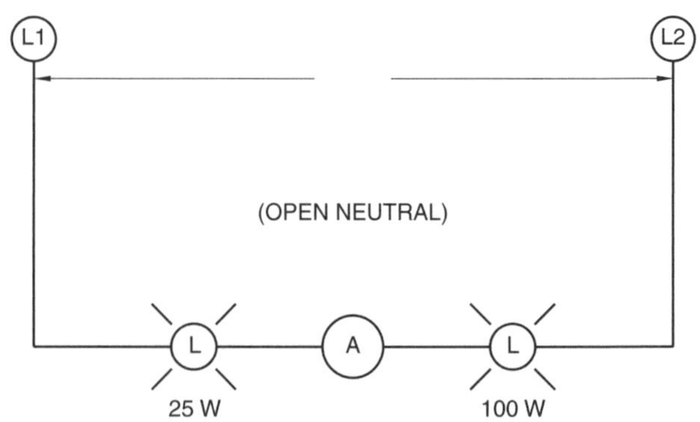

CIRCUIT WITH FAULT

CHAPTER 2

Electrical Formulas

Review Questions 2-4

Name ______________________ **Date** ______________

The answers for all review questions are listed in the answer key located in the back of this book.

______ **1.** A 1 kVA, 1ϕ transformer has a 120 V primary drawing 8.33 A and a 240 V secondary drawing 3.9 A. The transformer is ___% efficient.

A. 100 C. 94
B. 85 D. none of the above

______ **2.** A 7½ HP, 240 V, 1ϕ motor has an FLA of 40 A. The motor is ___% efficient.

A. 70 C. 45
B. 85 D. 58

______ **3.** An electrical circuit has 10 A of current and a resistance of 58 Ω. The voltage is ___ V.

A. 48 C. 580
B. 68 D. none of the above

______ **4.** A 3 AWG THW Cu, 3-wire, 120/240 V, 1ϕ feeder is 219′ long. The amperage available per phase is ___ A. (Use approximate K and anticipate a maximum recommended voltage drop)

A. 100 C. 75
B. 80 D. 67

______ **5.** A 4 Ω, 8 Ω, and 15 Ω resistor are connected in parallel. The total resistance is ___ Ω.

A. 27 C. 8.6
B. 3.25 D. 2.27

______ **6.** A 4 Ω resistor is connected in series with a 24 Ω and a 1 Ω resistor. The voltage drop across the 1 Ω resistor is 5 V. The supply voltage is ___ V.

A. 145 C. 5
B. 120 D. 105

______ **7.** The power needed for an 80% efficient, 5 HP, 240 V, 1ϕ motor is ___ VA.

A. 3648 C. 2884
B. 3730 D. 4662

______ **8.** A 10 kW, 240 V, 1ϕ load has an 85% power factor. The amperage drawn is ___ A.

A. 41.6 C. 39.2
B. 49 D. 118

______ **9.** A 15 kVA, 1ϕ load has a 240 V-rated source of supply. The amperage drawn is ___ A.

A. 6.25 C. 50
B. 68.1 D. 62.5

________ **10.** A 21 A, 230 V, 1ϕ circuit with a power factor of 93% has ___ VA.

A. 4830 C. 5193
B. 213.9 D. 4492

________ **11.** A 220 V motor operating for ¾ of an hour has a current of 10 A. A total of ___ Wh are used.

A. 2933 C. 2200
B. 1650 D. 75

________ **12.** Electricity cost 12¢ per kWh, how much will it cost to operate the motor in Problem 11?

A. $19.80 C. $0.09
B. $0.20 D. $35.40

________ **13.** A 1ϕ transformer has a 120 V primary and a 12 V secondary. The transformer has a rated output of 0.5 kW and is 95% efficient. The primary power is ___ VA.

A. 526 C. 475
B. 500 D. none of the above

________ **14.** A 120 V 1ϕ heating element rated at 2200 W draws ___ A.

A. 18.3 C. 20
B. 1.83 D. 15

________ **15.** ___ law states that the current in a circuit is proportional to the voltage and inversely proportional to the resistance.

A. Lenz's C. Kirchhoff's
B. Ohm's D. Newton's

________ **16.** A 10 Ω resistor has a voltage drop of 80 V and is connected in series with an unknown resistor. The supply voltage is 120 V. The resistance of the unknown resistor is ___ Ω.

A. 5 C. 40
B. 1.5 D. 8

________ **17.** One conductor of a 120 V circuit for a 6.5 Ω heater has a resistance of 0.193 Ω. The total resistance of the complete circuit is ___ Ω.

A. 6.5 C. 6.88
B. 6.69 D. 12

________ **18.** Four 4 Ω heaters are connected in parallel. The total resistance is ___ Ω.

A. 16 C. 1
B. 4 D. none of the above

________ **19.** True power equals apparent power when the power factor is ___.

A. leading C. 100%
B. lagging D. zero

________ **20.** Opposition to the flow of electrons is ___.

A. ohms C. resistance
B. voltage D. current

Learner Resources
atplearningresources.com/quicklinks
Access Code: 533920

CHAPTER 3

NEC® Examples

This chapter contains examples of how the NEC® is applied. Each of the questions and problems in this chapter is an example of applying the NEC® to solve electrical problems. The NEC® references should be thoroughly studied.

CONDUIT FILL

NEC® Chapter 9, Tables contains requirements and mandatory rules for conduit fill. The various tables contain information to safely calculate conduit fill. Table 1 contains the percent fill of a conduit or raceway for a set amount of conductors. Chapter 9, Table 4 lists the trade sizes in standard inches and millimeters, internal diameter in inches, total area 100% sq in., and the amount of fill in sq in. based on the amount of conductors installed. This information is provided for different types of conduit and tubing.

Chapter 9, Table 5 lists the dimension of insulated conductors and fixture wire. This table gives the type, size, approximate diameter, and approximate area in sq in. *Note:* Pay close attention to *. Chapter 9, Table 5A covers compact copper and aluminum conductors. A *compact conductor* is a conductor which has been compressed to eliminate voids between strands.

Informative Annex C provides the maximum amount of conductors and fixtures wires that are permitted to be installed in different types of conduit or tubing. Tables C1 to C13 cover conductor and fixture wires of the same size and type of insulation for trade sizes from ⅜″ to 6″ conduit and tubing. *Note:* Pay close attention to *. Tables C1(A) through C13(A) cover compact conductors installed in different types of conduits or tubing. The tables are for conduit and tubing sizes when all conductors or fixture wires are the same size and have like insulation.

Examples: Conduit Fill

1. How many 12 AWG THW Cu conductors can be installed in a ¾″ EMT conduit?

Informative Annex, Table C1: ¾″ = 11
Conductors = **11**

2. What size RMC is required for three 12 AWG THW Cu, three 12 AWG TW Cu, three 12 AWG RHW Cu (w/o outer covering), and three 8 AWG THWN Cu conductors?

Ch 9, Table 5:
Three 12 AWG THW Cu   0.0181 × 3 = 0.0543
Three 12 AWG TW Cu   0.0181 × 3 = 0.0543
Three 12 AWG RHW Cu (w/o outer covering)
  0.026 × 3 = 0.078
Three 8 AWG THWN Cu   0.0366 × 3 = 0.1098
  0.2964

Ch 9, Table 1; Ch 9, Table 4 (RMC):
Over 2 conductors = 40% fill
1″ conduit = 0.355
RMC = **1″**

3. How many 12 AWG THHN Cu conductors can be installed in a ½″ EMT conduit 23″ long?

Ch 9, Note 4: Less than 24″ = 60% fill
Ch 9, Table 4: 100% fill = 0.304
0.304 × 0.60 = 0.1824

Ch 9, Table 5: THHN Cu = 0.0133
0.1824 ÷ 0.0133 = 13.7
Ch 9, Note 7: Drop 0.7
Conductors = **13**

4. How many 8 AWG TW Cu conductors can be installed in a ¾″ IMC conduit 12″ long?

Ch 9, Note 4: Less than 24″ = 60% fill
Ch 9, Table 4: 100% fill = 0.586
0.586 × 0.60 = 0.3516
Ch 9, Table 5: 8 AWG TW Cu = 0.0437
0.3516 ÷ 0.0437 = 8
Conductors = **8**

5. What size rigid PVC, Schedule 40 is required for three 8 AWG TW Cu phase conductors, one 10 AWG TW Cu EGC, and one 8 AWG stranded bare Cu conductor?

Ch 9, Table 5: Three 8 AWG TW Cu 0.0437 × 3 = 0.1311
One 10 AWG TW Cu 0.0243 × 1 = 0.0243
Ch 9, Table 8:
One 8 AWG bare Cu 0.017 × 1 = 0.017
0.1724
Ch 9, Table 1; Ch 9, Table 4: ¾″ = 0.203
PVC = **¾″**

6. How many 6 AWG THW compact Al conductors can be installed in a 1¼″ flexible metal conduit?

Informative Annex C, Table C 3(A): 1¼″ = 7
Conductors = **7**

7. What size EMT is required for three 1/0 AWG and one 1 AWG XHHW Al compact conductors?

Ch 9, Table 5A:
Three 1/0 AWG XHHW Al 0.1590 × 3 = 0.477
One 1 AWG XHHW Al 0.1352 × 1 = 0.1352
0.6122
Ch 9, Table 1; Ch 9, Table 4: 1½″ = 0.814
EMT = **1½″**

CONDUCTOR ADJUSTMENT FACTOR

The ampacities listed in Table 310.16 are based on three current-carrying conductors in a raceway. Chapter 9, Table C1 allows nine 12 AWG THWN conductors in a ½″ raceway. Section 310.15 states that the allowable ampacity shall be reduced per the percentages listed in Table 310.15(C)(1). When the number of current-carrying conductors exceeds three, the overcurrent protection shall be adjusted to reduce the allowable ampacity of the conductors. When the adjusted ampacity does not correspond to a standard rating under 800 A, the next higher standard rating is permitted. See 240.6.

The grounded (neutral) conductor that caries only unbalanced current shall not be counted for adjustment per 310.15(E)(1). For a 3ϕ, 4-wire, wye system using two phases of the phase conductors, the neutral is considered a current-carrying conductor per 310.15(E)(2). For a 3ϕ, 4-wire, wye system where the major portion of the load consists of nonlinear loads such as electric discharge lighting (ballast operated), the neutral is considered a current-carrying conductor per 310.15(E)(3).

Examples: Conductor Adjustment

8. What is the ampacity of eight 8 AWG THW 75°C Cu current-carrying conductors installed in a raceway?

Table 310.16: 8 AWG THW 75°C = 50 A
Table 310.15 (C)(1): 50 A × 0.70 = 35 A
Ampacity = **35 A**

9. A piece of electrical equipment draws 35 A. The raceway contains eight current-carrying conductors. What size THW 75°C Cu conductors are required?

Table 310.15(C)(1):

$$\frac{1}{0.70} = 1.428 \text{ (multiplier)}$$

1.428 × 35 A = 49.98 A
Conductors = **8 AWG THW Cu rated at 50 A**

10. What is the ampacity of four 12 AWG THWN 75°C Cu current-carrying conductors installed in a 23″ raceway?

Table 310.16; 310.15(C)(1)(a)
12 AWG THWN 75°C Cu = 25 A
Ampacity = **25 A**

11. What is the ampacity of three 10 AWG THW 75°C Cu current-carrying conductors and one 10 AWG THW 75°C Cu EGC?

Table 310.16; 310.15(F):
10 AWG THWN 75°C Cu = 35 A
Ampacity = **35 A**

12. What is the ampacity of four 12 AWG TW Cu conductors feeding three balanced resistive lighting circuits?

Table 310.16; 310.15(E)(1):
12 AWG TW Cu = 20 A
Ampacity = **20 A**

13. A raceway contains four conductors and feeds three 120 V fluorescent lighting circuits utilizing a shared neutral on a 208Y/120 V system. What is the allowable ampacity of the 12 AWG THWN conductors?

Table 310.16; Table 310.15(C)(1); 310.15(E)(2):
12 AWG TW Cu = 25 A
25 A × 0.80 = 20 A
Ampacity = **20 A**

14. What is the ampacity of three incandescent lighting circuits fed from a 3ϕ, 4-wire, wye system using 12 AWG THW 75°C Cu conductors?

Table 310.16; 310.15(E)(1):
12 AWG TW Cu = 25 A
25 A × 0.80 = 20 A
Ampacity = **20 A**

15. Five current-carrying 6 AWG TW 60°C Cu conductors are required for resistive loads. What is the ampacity of the conductors and the maximum overcurrent protection permitted?

Table 310.16:
6 AWG TW Cu = 55 A
55 A × 0.80 = 44 A
Ampacity = **44 A**
240.4(B)(2) and (3); 240.6(A): Next higher size = 45 A
OCPD = **45 A**

TEMPERATURE CORRECTION

A conductor shall not be used in such a manner that its operating temperature exceeds that designated for the type of insulation per 310.15(B)(2). Ampacities are based on an ambient temperature of 30°C (86°F). Ambient temperature is the temperature surrounding a device. Correction factors are applied for other ambient temperatures based on 30°C (86°F) per Table 310.15(B)(1) or for other ambient temperatures based on 40°C (104°F) per Table 310.15(B)(2).

Examples: Temperature Correction

Note: The following questions refer to correction factors for ambient temperatures based on 30°C (86°F).

16. What is the ampacity of a 3-wire circuit using THW 75°C, 8 AWG Cu conductors in an ambient temperature of 113°F?

Table 310.16: 8 AWG THW 75°C Cu = 50 A
Table 310.15(B)(1), Correction Factors: 113°F = 0.82
0.82 × 50 = 41 A
Ampacity = **41 A**

17. What size THW 75°C Cu conductors are required for a 3-wire circuit with a load of 41 A in an ambient temperature of 113°F?

Table 310.15(B)(1), Correction Factors: 113°F = 0.82

$$\frac{1}{0.82} = 1.2195 \text{ (multiplier)}$$

41 A × 1.2195 = 49.9 A
Conductors = **8 AWG THW 75°C Cu rated at 50 A**

18. What is the ampacity of three 10 AWG TW 60°C Cu conductors in an ambient temperature of 57°C?

Table 310.16: 10 AWG TW 60°C Cu = 30 A
Table 310.15(B)(1), Correction Factors: Not listed for greater than 55°C for insulated conductor with temperature rating of 60°
Ampacity = **Not Permitted** (due to degradation of TW insulation)

SINGLE MOTORS 1ϕ (GENERAL DUTY)

Five factors that concern single motors are full-load current, conductor size, overloads, overcurrent protection, and disconnecting means.

Full-Load Current (FLC)

The FLC for general duty motors shall be based on the HP and voltage ratings listed in Tables 430.248 through 430.250 per 430.6(A)(1). The FLC shall not be based on the motor's nameplate full-load current rating.

Example: Full-Load Current

19. What is the FLC of a 3 HP, 230 V, 1ϕ motor?

Table 430.248: 3 HP, 230 V, 1ϕ = 17 A
FLC = **17 A**

Conductor Size

Branch-circuit conductors supplying a single motor shall have an ampacity not less than 125% of the motor full-load current rating per 430.22.

Example: Conductor Size

20. What is the minimum size THW 75°C Cu conductors required for a 3 HP, 230 V, 1ϕ motor?

Table 430.248: 3 HP, 230 V, 1ϕ = 17 A
430.22: 17 A × 125% = 21.25 A
Table 310.16: THW 75°C Cu = 25 A
Conductors = **12 AWG THW 75°Cu**

Overloads

A separate motor overload protection shall be based on the motor's nameplate full-load current rating per 430.6. *Overloads* are heat-sensing devices intended to protect the motor from excessive current. Short circuits are not considered an overload. *Note:* Testing authority questions may not provide the nameplate rating for a motor. Therefore, use the FLC ratings listed in the Tables in 430 for overload calculations.

Examples: Overloads

21. In general, what is the maximum overload rating for a 3 HP, 230 V, 1ϕ motor?

Table 430.248: 3 HP, 230 V, 1ϕ = 17 A
430.32(A)(1): 17 A × 115% = 19.55 A
Maximum Overload Rating = **19.55 A**

22. A 3 HP, 230 V, 1ϕ motor has internal thermal protection. What is the maximum trip current rating?

Table 430.248: 3 HP, 230 V, 1ϕ = 17 A:
430.32(A)(2): 17 A × 156% = 26.52 A
Maximum Trip Current = **26.52 A**

23. A 3 HP, 230 V, 1ϕ motor has overloads sized per 430.32(A)(1). The motor starts, but fails to carry the load. What is the maximum overload permitted?

Table 430.248: 3 HP, 230 V, 1ϕ = 17 A
430.32(C): 17 A × 130% = 22.1 A
Maximum Overload = **22.1 A**

Overcurrent Protection

The starting current for a motor can be from 2 to 10 times the FLC, therefore the overcurrent protection must be sized to allow the motor to start. The overcurrent protection device is for protection against short circuits and ground-faults. The overcurrent protection device shall be capable of carrying the starting current per 430.52. When ratings in Table 430.52 are not sufficient for the starting current of the motor, the percentages listed in 430.52(C)(1), Exception No. 2(1), (2), (3), and (4) may be applied.

Examples: Overcurrent Protection

24. What is the maximum size ITCB permitted when using a 3 HP, 230 V, 1ϕ motor with no code letter?

Table 430.248: 3 HP, 230 V, 1ϕ = 17 A
Table 430.52: ITCB = 250%
17 A × 250% = 42.5 A
430.52(C)(1), Ex. 1: 42.5 A
240.6: Next higher size = 45 A
Maximum Size ITCB = **45 A**

25. A 3 HP, 230 V, 1ϕ motor with no code letter has TDFs that are not sufficient for the starting current. What is the maximum size TDFs permitted?

Table 430.248: 3 HP, 230 V, 1ϕ = 17 A
430.52(C)(1), Ex. 2(2): 17 A × 225% = 38.25 A
Maximum Size TDFs = **35 A**

26. A 3 HP, 230 V, 1ϕ motor with no code letter has an ITCB that is not sufficient for starting the motor. What is the maximum size ITCB permitted?

Table 430.248: 3 HP, 230 V, 1ϕ = 17 A
430.52(C)(1), Ex. 2(3): 17 A × 400% = 68 A
Maximum Size ITCB = **60 A**

Disconnecting Means

A *disconnecting means* is a device or group of devices that separate or isolate the conductors of a circuit from their source of supply. They may, or may not, provide overcurrent protection (either fuses or CBs). The disconnecting means is the point at which personnel may turn off, lockout, and tagout power prior to maintenance. Disconnecting means terminal points are excellent to take voltage and current measurements. Disconnecting means are covered in 430, Part IX.

Example: Disconnecting Means

27. What is the minimum ampere rating for the disconnecting means for a 3 HP, 230 V, 1ϕ motor?

Table 430.248: 3 HP, 230 V, 1ϕ = 17 A
430.110(A): 17 A × 115% = 19.55 A
Minimum Ampere Rating = **19.55 A**

MOTOR FEEDERS

Conductors supplying several motors shall have ampacity at least equal to the sum of the FLC rating of all the motors, plus 25% of the highest-rated motor in the group per 430.24. The feeder protection device shall be based on the largest branch-circuit OCPD of the group plus the FLC ratings of the other motors per 430.62(A).

Examples: Motor Feeders

28. A 230 V, 1ϕ feeder is to supply 1 HP, 5 HP, and 7½ HP motors. What are the minimum size conductors using THW 75°C Cu?

Table 430.248:
1 HP FLC = 8 A
5 HP FLC = 28 A
7½ HP FLC = 40 A
430.24: 40 × 125% = 50 + 28 + 8 = 86 A
Table 310.16: 3 AWG THW 75°C Cu = 100 A
Minimum Size Conductors = **100 A**

29. What is the minimum size ITCB required for feeder overcurrent protection in Problem 28?

Table 430.248: 7½ HP = 40 A
Table 430.52: 40 A × 250% = 100 A (standard size)
430.62: 100 A (largest OCPD of the group) + 28 A + 8 A = 136 A
240.6: Next smaller size = 125 A
Minimum Size ITCB = **125 A**

30. A 3-wire feeder supplies two 1 HP, 120 V motors and one 3 HP, 230 V motor. What are the minimum size conductors for the feeder using THW 75°C Cu?

Table 430.24: 3 HP, 230 V = 17 A × 125% = 21.25 A
1 HP, 120 V = 16 A

L1	N	L2
21 A	0	21 A
16 A	0	16 A
37 A	0	37 A

Table 310.16: 8 AWG THW 75°C Cu = 50 A
Minimum Size Conductors = **50 A**

31. What is the maximum size TDFs required in Problem 30?

Table 430.52: 3 HP, 230 V = 17 A × 175% = 29.75 A
240.6: Next higher size = 30 A
30 A + 16 A = 46 A
240.6: Next lower size = 45 A
Maximum Size TDFs = **45 A**

RANGES (HOUSEHOLD)

Table 220.55 lists the demand factors for ranges. Actual load and demand load are very different. Demand factors are based on the diversified use of the range, due to the fact that all parts of the range are not used at the same time.

Table 220.55, Column C is used for ranges that have a nameplate rating from 8¾ kW to 12 kW. For a range with a nameplate rating that falls between 8¾ kW and 12 kW, the demand load shall be computed at 8 kW. In calculations when the 12 kW rating is exceeded, refer to Note 1.

Table 220.55, Column A is used for ranges with a nameplate rating less than 3½ kW. The nameplate rating is multiplied times the percentages listed in the Column to calculate the demand load.

Table 220.55, Column B is used for ranges with a nameplate rating that falls between 3½ kW and 8¾ kW. The nameplate rating is multiplied times the percentage listed in the Column to calculate the demand load.

Table 220.55, Note 1 states that the demand rating listed in Table 220.55, Column C shall be increased by 5% for every kW above 12 kW.

Table 220.55, Note 2 states that for ranges over 8¾ kW through 27 kW of different ratings, find the average range and apply Column C, increasing the values for every kilowatt by 5% when 12 kW is exceeded. Use 12 kW for any range rated less than 12 kW.

Table 220.55, Note 3 grants permission to add the nameplate ratings of ranges that fall between 1¾ kW to 8¾ kW and multiply the total by the percentages listed in Column A or Column B for the number of appliances.

Table 220.55, Note 4 permits the branch-circuit load for a range to be based on Table 220.55. No demand is permitted for one wall-mounted oven or one counter-mounted cooking unit. The branch-circuit load shall be based on the nameplate rating of the unit.

One counter-mounted cooking unit and not more than two wall-mounted ovens are permitted to be treated as one range. Add the nameplate ratings of the units and use Column C.

Table 220.55, Note 5 permits the use of Table 220.55 for load calculations of household ranges used in a home-cooking class in a school.

For ranges 8¾ kW or more, the minimum branch-circuit rating shall be 40 A per 210.19(A)(3). For ranges 8¾ kW or higher, the ampacity of the neutral conductor shall not be less than 70%. The neutral conductor shall not be smaller than 10 AWG per 210.19(A)(3), Exception No. 2.

The maximum unbalanced load of a feeder supplying a household range shall be considered as 70% of the load on the ungrounded conductors per 220.61(B).

Examples: Ranges

32. What is the demand load for one 10 kW range?

Table 220.55, Col C: 8 kW

Demand Load = **8 kW**

33. What is the demand load for a 3 kW range?

Table 220.55, Col A:

3000 W × 80% = 2400 W = 2.4 kW

Demand Load = **2.4 kW**

34. What is the demand load for an 8 kW range?

Table 220.55, Col B:

8000 W × 80% = 6400 W = 6.4 kW

Demand Load = **6.4 kW**

35. What is the demand load for a 13 kW range?

Table 220.55, Col C: 13 kW – 12 kW = 1 kW

Table 220.55, Note 1:

8000 W × 105% = 8400 W = 8.4 kW

Demand Load = **8.4 kW**

36. What is the demand load for five 3 kW ranges?

Table 220.55, Col A:

5 × 3000 W × 62% = 9300 W = 9.3 kW

Demand Load = **9.3 kW**

37. What is the demand load for five 7 kW ranges?

Table 220.55, Col B:

5 × 7000 W × 45% = 15,750 W = 15.75 kW

Demand Load = **15.75 kW**

38. What is the demand load for five 9 kW ranges?

Table 220.55, Col C: 20 kW

Demand Load = **20 kW**

39. What is the demand load for five 15 kW ranges?

Table 220.55, Col C: 15 kW – 12 kW = 3 kW

Table 220.55, Note 1:

20,000 W × 115% = 23,000 W = 23 kW

Demand Load = **23 kW**

40. A 9 kW, 240 V range is to be wired using TW Cu conductors. What are the minimum sizes permitted for the ungrounded conductors, grounded conductors, and the overcurrent protection?

Ungrounded Conductors

Table 220.55, Col. C: 8000 kW 240 V = 33.3 A

Table 310.16: 33.3 A = 8 AWG TW Cu

Ungrounded Conductors = **8 AWG TW Cu**

Grounded Conductors

210.19(A)(3), Ex. 2: 10 AWG TW Cu

Grounded Conductors = **10 AWG TW Cu**

OCPD

210.19(A)(3): 33.3 A requires 40 A

OCPD = **40 A**

41. What branch-circuit current rating is required for a 240 V, 12 kW range?

Table 220.55, Col C; Table 220.55, Note 4:

$$\frac{8000\text{ W}}{240\text{ V}} = 33.3\text{ A (210.19(A)(3))}$$

Branch-Circuit Current Rating = **40 A**

42. What branch-circuit current rating is required for one 6 kW countertop unit and two 4 kW wall ovens are wired on the same 240 V branch-circuit?

Table 220.55, Note 4:

6 kW + 4 kW + 4 kW = 14 kW

Table 220.55, Col C; 210.19(A)(3):

8 kW × 110% = 8800 W/240 V = 36.6 A

Branch-Circuit Current Rating = **40 A**

43. What is the neutral load for five 12 kW ranges?

Table 220.55 Col C: Five 12 kW ranges = 20 kW

220.61(B): 20 kW × 70% = 14 kW

Neutral Load = **14 kW**

44. What is the demand load for five 10 kW ranges, five 12 kW ranges, and five 14 kW ranges?

Table 220.55, Col C; Table 220.55, Note 2:

5 × 12 = 60

5 × 12 = 60

<u>5 × 14 = 70</u>

15 ranges = 190 kW

190 kW ÷ 15 = 12.6 kW (0.6 major fraction)

Table 220.55, Col C: 15 ranges = 30 kW

30 kW × 1.05 = 31.5 kW (0.5 is not permitted to be dropped)

Demand Load = **31.5 kW**

CALCULATING BOX FILL FOR CONDUCTORS

Table 314.16(A) lists standard metal boxes. Standard metal boxes are not required to be marked with a cubic inch capacity. Boxes 100 cu in. or less not listed in the table, including plastic boxes, are required to be marked with their cubic inch capacity. Table 314.16(A) lists the minimum cubic inch capacity and the maximum number of the same size conductors permitted to be installed for the different types and sizes of boxes. The total volume of a box is based on the total volume of assembled sections, such as plaster rings, box extensions, etc. Table 314.16(A) is based on an empty box with no assembled sections. Deductions from the conductor count listed in Table 314.16(A) are required based on 314.16(B). These deductions include the following:

- Each ungrounded and grounded conductor requires one deduction. An unbroken conductor that passes through a box requires one deduction. Each unbroken loop longer than 12″ requires two deductions.
- A box containing one or more cable clamps requires one deduction.
- A box containing a hickey or a fixture stud requires one deduction.
- Each yoke or strap with one or more devices requires two deductions. For example, an S/P switch or a duplex receptacle requires a deduction of two.
- The total amount of all equipment-grounding conductors and bonding jumpers requires one deduction.

Per 314.16(B)(1), Exception, an equipment-grounding conductor and not over four luminaire (fixture) wires smaller than 14 AWG are permitted to be omitted from the calculations.

Table 314.16(B) is used for calculating box fill with different size conductors. The same deductions (A through F) including the exception apply. The deductions must be based on the largest conductor size entering the box.

Examples: Box Fill

45. How many 12 AWG conductors are permitted in a 4″ sq 1½″ box?

Table 314.16 (A): **nine conductors**

46. How many 14 AWG conductors may be added to a device box 3″ × 2″ × 3½″ that contains an S/P switch, two cable clamps, and a 14/2 w/grd Type MN cable?

Deductions

S/P switch: 2

Two cable clamps: 1

Two 14 AWG conductors: 2

Grounding conductor: 1

Total deductions: 6

Table 314.16 (A): 3″ × 2″ × 3½″ device box = nine 14 AWG conductors

9 − 6 = 3

Three 14 AWG conductors may be added.

47. What is the minimum cubic inch volume required for a box that contains three 12 AWG conductors and one 12 AWG equipment-grounding conductor, and four 14 AWG and three 10 AWG conductors with one equipment-grounding conductor?

Table 314.16(B): (12 AWG) 3 × 2.25 = 6.75; (14 AWG) 4 × 2.00 = 8; (10 AWG) 3 × 2.50 = 7.5; equipment ground = 2.5

6.75 + 8 + 7.5 + 2.5 = 24.75

Total volume required = **24.75 cu in.**

48. What is the minimum cubic inch volume required for a box that contains two cable clamps, one 14/2 w/grd Type NM cable, one 10/2 w/grd Type NM cable, a duplex receptacle, and three 16 AWG fixture wires?

Table 314.16(B): (cable clamps) 1 × 2.5 = 2.5; (14 AWG conductors) 2 × 2 = 4; (10 AWG conductors) 2 × 2.5 = 5; (receptacles) 2 × 2.5 = 5; (equipment grounds) 1 × 2.5 = 2.5; fixture wire (omit) = 0

2.5 + 4 + 5 + 5 + 2.5 = 19

Total volume required = **19 cu in.**

CONDUCTOR RESISTANCE

Chapter 9, Table 8 is used to find the resistance values of conductors. *Resistance* is opposition to the flow of electrons. The unit of measurement for resistance is the Ohm (Ω). Electrical conductors are expressed in circular mils or AWG. The American wire gauge (AWG) is a standard for wire sizes. Four factors that affect the resistance of conductors are:

- Size: Cross-sectional area measured in circular mils (CM)
- Type of Material: Cu or Al

- Length: Ω/kFT
- Temperature: DC resistance at 75°C (167°F)

Size. The size of a conductor affects the resistance. The resistance of a wire is inversely proportional to its cross-sectional area. For example, for a ¼″ OD conductor, the resistance is four times greater than the resistance of a ½″ OD conductor. As the conductor size increases, the resistance decreases.

Type of Material. The type of material of a conductor affects the resistance. If the type of material for conductors with the same cross-sectional area and the same length differ, the resistance also differs. For example, the resistance for 1000′ of 12 AWG Cu conductor is 1.93 Ω. The resistance for 1000′ of 12 AWG Al conductor is 3.18 Ω. See Chapter 9, Table 8.

Length. The length of a conductor affects the resistance. The shorter the conductor, the less the resistance. The longer the conductor, the more the resistance. For example, the resistance for 1000′ of 12 AWG Cu conductor is 1.93 Ω. The resistance for 2000′ of 12 AWG Cu conductor is 3.86 Ω.

Temperature. The temperature of a conductor affects the resistance. The resistance of all conductors increase as the temperature of the conductor increases. The resistance values of Chapter 9, Table 8 are based on 75°C (167°F).

Examples: Conductor Resistance

49. What is the area in circular mils of a 14 AWG Cu, solid, uncoated, bare conductor?

Ch 9, Table 8: 4110 CM

Area = **4110 CM**

50. What is the diameter of a 14 AWG Cu, solid, uncoated, bare conductor?

Ch 9, Table 8: 0.064″

Diameter = **0.064″**

51. What is the resistance of a 14 AWG Cu, solid, uncoated, bare conductor per 1000′?

Ch 9, Table 8: 3.07 Ω

Resistance = **3.07 Ω**

52. What is the cross-sectional area of a 14 AWG Cu, solid, uncoated, bare conductor in sq in.?

Ch 9, Table 8: 0.003 sq in.

Area = **0.003 sq in.**

53. What is the resistance of 125′ of solid, 10 AWG THW Al wire?

Ch 9, Table 8:

$$10 \text{ AWG } \Omega/\text{kFT} = \frac{2}{1000} = 0.002\ \Omega/\text{ft}$$

$$0.002 \times 125' = 0.25\ \Omega$$

Resistance = **0.25 Ω**

54. What is the length of solid 8 AWG Cu with a resistance of 0.24 Ω?

Ch 9, Table 8:

$$8 \text{ AWG } \Omega/\text{kFT} = \frac{0.764}{1000} = 0.000764\ \Omega/\text{ft}$$

$$\frac{0.24}{0.000764} = 314'$$

Length = **314′**

55. What is the total resistance per 1000′ of a parallel set of 2/0 AWG Cu conductors?

Ch 9, Table 8:

$$2/0 \text{ AWG } \Omega/\text{kFT} = 0.0967$$

$$R_T = \frac{R_1}{N} = \frac{0.0967}{2} = 0.04835\ \Omega$$

Total Resistance = **0.04835 Ω**

DERATING CONDUCTORS INSTALLED ON ROOFTOPS

Conductors installed on or above rooftops shall be derated when exposed to direct sunlight. Whenever possible, raceways or cables shall be installed with a minimum clearance of ⅞″ from the bottom of the raceway or cable to the top of the roof. If the clearance from the bottom of the raceway or cable is less than ⅞″, then a temperature adder of 33°C (60°F) must be added to the outdoor temperature to determine the appropriate ambient temperature correction factor for the application, based on the values listed in either Table 310.15(B)(1) or Table 310.15(B)(2). The AHJ will determine the average ambient temperature in the area based on the American Society of Heating, Refrigerating and Air-Conditioning Engineers' (ASHRAE) average listings of ambient temperatures in various locations.

Examples: Derating Conductors Installed on Rooftops

56. Calculate the ampacity for two 8 AWG THWN-2 Cu conductors installed in EMT raceway located ½″ above a rooftop in Tampa, Florida. Per ASHRAE, the average rooftop temperature for Tampa is 92°F.

310.15(B)(2): Conductors less than ⅞″ above the rooftop require a 60°F adder; 92 + 60 = 152°F

Table 310.15(B)(1): Ambient temperature of 152°F using the 90°C column = 0.58 correction factor

Table 310.16: 8 AWG THWN-2 Cu Conductors = 55 A

55 A × 0.58 = 32 A

Ampacity = **32 A**

57. Determine the conductor size for THWN-2 Cu conductors supplying a 36 A load. The conductors are installed in EMT located 12″ above a rooftop in Tampa, Florida. The temperature inside a conduit in direct sunlight is 122°F.

Table: 310.15(B)(1): Ambient temperature of 122°F using the 90°C column = 0.82 correction factor

Use a multiplier 1 ÷ 0.82 = 1.22

1.22 × 36 A = 44 A

Table 310.16: THWN-2 Cu conductors with an allowable ampacity of 44 A = 8 AWG conductors

Conductor size = **8 AWG THWN-2 Cu**

ELECTRIC WELDERS

Article 630 covers individual welders. Subsections 630.11(A) and (B) cover ampacity for supply conductors for arc welders.

Supply conductors shall not be more than the I_{1eff} value on the welder nameplate. If no I_{1eff} is listed, the ampacity of the supply conductors shall be determined by multiplying the rated primary current in amperes listed on the nameplate and the duty cycle multiplier for the arc welder.

Overcurrent protection shall be rated at not more than 200% of the I_{1max} (the maximum amount of rated supply current at the maximum rated output) or the rated primary current of the arc welder. Section 630.12 permits using the next higher standard rating (240.6).

Examples: Electric Welders

58. An AC and DC TIG or stick welder has a rated output of 300 A, 30 V, with a 60% duty cycle. The input current is 91 A with a 208 V source of supply. What is the ampacity required for the supply conductors?

Subsection 630.11(A)

Primary current = 91 A

Duty cycle = 60%

Multiplier = 0.78

91 A × 0.78 = 70.98 A

Ampacity of supply conductors = **71 A**

59. A DC motor-generator arc welder has a 230 V, 3ϕ, 78 A input. The output is 400 A/36 V with a 60% duty cycle. What is the ampacity required for the supply conductors?

Subsection 630.11(A)

Primary current = 78 A

Duty cycle = 60%

Multiplier = 0.81

78 A × 0.81 = 63.18 A

Ampacity of supply conductors = **63 A**

60. A resistance spot welder has a rated output of 10 kVA with a 50% duty cycle. The input rating is 220 V/45 A. What is the ampacity required for the supply conductors?

Subsections 630.31(A)(1) and (2)

Primary current = 45 A

Duty cycle = 50%

Multiplier = 0.71

45 A × 0.71 = 31.95 A

Ampacity of supply conductors = **32 A**

61. A DC welder with capability of stick and TIG welding in addition to air carbon arc gouging has an output of 250 A/30 V with a 50% duty cycle. The machine has a 460 V, 1ϕ supply drawing 33 A. What is the maximum rating of the OCPD?

Subsection 630.12(A)

33 A × 200% = 66 A

Cannot exceed 200%. Next lower standard rating shall be used per 240.6(A)

Maximum OCPD rating = **60 A**

SERVICE CALCULATIONS (RESIDENTIAL)

Article 310.15(B)(7) permits a reduction in the size of the ungrounded service entrance conductors for a residential dwelling unit when the nominal voltage is 120/240 V and 100 A to 400 A. This article applies to 1ϕ, 120/240 V systems and 120/208 V systems. The grounded conductor (neutral) is permitted to be smaller than the ungrounded conductor. For a 120/208 V system, no reduction is permitted in the grounded conductor per 220.61(C)(1).

Examples: Service Calculations

62. A dwelling has a 200 A, 120/240 V, 1ϕ service with THW 75°C Cu conductors. What is the size of the ungrounded and grounded conductors?

Ungrounded Conductors

310.12: 200 A × 0.83 = **166 A**

Table 310.16: 120 V THW 75°C Cu conductors = **2/0**

Grounded Conductor

Table 250.102(C)(1): 2/0 or 3/0 Cu = 4 AWG grounded conductor = **4 AWG Cu**

Article 310.12 permits the grounded conductor to be smaller than the ungrounded conductor. The AHJ may permit a dropping of two trade sizes or may require a load calculation to reduce the neutral size.

63. An apartment building has a 120/208 V, 3ϕ, 4-wire service with a 100 A, 120/208 V, 1ϕ feeder of THW Cu to each apartment. What are the sizes for the ungrounded and grounded conductor to each apartment?

Ungrounded Conductor

310.12: 100 A × 0.83 = 83 A

Ungrounded Conductor = **4 AWG**

Grounded Conductor

310.15(E)(2): No reduction permitted for the neutral

Grounded Conductor = **4 AWG Cu**

Article 310.15(E)(2) states that the neutral conductor, when using two phases of a 3ϕ, 4-wire, wye system, carries approximately the same current as the other conductors.

64. The GEC in Problem 62 is an underground water pipe. What size Cu GEC is required?

250.66(A): Not required to be larger than 6 AWG Cu

GEC = **6 AWG Cu**

The GEC is based on the largest service entrance conductor per Table 250.66.

65. The grounding electrode in Problem 62 is a concrete-encased electrode. What size Cu GEC is required?

250.52(A)(3);250.66(B): Not required to be larger than 4 AWG Cu

GEC = **4 AWG Cu**

SERVICE LOAD CALCULATIONS (OPTIONAL CALCULATION–220.82)

Article 220 permits two methods for computing service entrance loads for a dwelling unit. Part I is general information. Parts III and IV are used for the standard and optional methods for calculating service entrance loads and conductor sizes. The Optional Calculation is a quick and simple method to use. It is the most common method used for computing service entrance loads for dwellings, although the loads may calculate slightly higher than with the Standard Calculation. See 220.82. Informative Annex D of the NEC® shows examples for each method.

Examples: Optional Calculation

Use the following for Problems 66 through 74:

A dwelling unit has a 50′ × 42′ floor area with a 20 sq ft attached garage. The dwelling unit has a 3 kW range, 4.5 kW hot water heater, 5 kW electric dryer, ⅓ HP food waste disposal, 1.2 kW dishwasher, and a 40 A, 240 V heat pump with 10 kW of supplemental heat. No system interlock. The service is 120/240 V, 1ϕ.

66. What is the general lighting load (Optional Calculation)?

220.82(B)(1): 50′ × 42′ × 3 VA = 6300 VA

General Lighting Load = **6300 VA**

Allow 3 VA per sq ft for general lighting and general-use receptacles per 220.82(B)(1). The floor area is computed from the outside dimensions of the dwelling. For dwelling units, the computed floor area does not include open porches, garages, or unused or unfinished spaces not adaptable for future use per 220.12.

67. How many 15 A, 2-wire circuits are required for general lighting (Optional Calculation)?

$$I = \frac{P}{E}$$

$$I = \frac{6300}{120} = 52.5$$

$$\frac{52.5}{15} = 3.5 \text{ (round to 4)}$$

Two-Wire Circuits = **4**

All general-use receptacle outlets of 20 A or less in dwelling units (except for small appliance and laundry outlets) shall be considered for general illumination. No additional load calculations shall be required for such outlets per 220.14(J).

68. What is the calculated load for the ungrounded (hot) conductors (Optional Calculation)? See 220.82 and Informative Annex D, Example D2(c).

220.82(B)(1): General Lighting:
50′ × 42′ × 3 VA/sq ft = 6300 VA
220.82(B)(2):
Small Appliance & Laundry:
1500 VA × 2 + 1500 VA = 4500 VA
220.82(B)(3):
Range = 13,000 VA
Hot Water Heater = 4500 VA
Dryer = 5000 VA
220.82(B)(4): Table 430.248:
Food Waste Disposer:
7.2 A × 120 V = 864 VA
220.82(B)(3):
Dishwasher = 1200 VA
Total = 35,364 VA

Applying Demand Factors
220.82(B): 100% of first 10 kVA = 10,000 VA
Over 10 kVA × 40%
35,364 VA – 10,000 VA = 25,364 VA
25,364 VA × 40% = 10,146 VA
10,000 VA + 10,146 VA = 20,146 VA

Demand Load Without A/C and Heat = 20,146 VA
220.82(C)(3): Heat Pump and Supplementary Heat have no system interlock, 100% nameplate rating:
Heat Pump = 40 A × 240 V = 9600 VA
Supplementary Heat = 10,000 VA
9600 VA + 10 kVA × 65% = 16,100 VA
Informative Annex D, Example D2(c)
Demand Load of Heat Pump with Supplementary Heat = 16,100 VA
36,246 VA

$$Load = \frac{36{,}246\ \text{VA}}{240\ \text{V}} = 151\ \text{A}$$

Calculated Load for Ungrounded Conductors = **151 A**

69. What is the minimum size service and service entrance conductors using THW Cu conductors (Optional Calculation)?

Calculated Load for Ungrounded Conductors = 151 A
Table 310.16
Minimum Size Service = **175 A**
Service Entrance Conductors
Ungrounded Conductor = **2/0 AWG THW Cu**
Grounded Conductor = **4 AWG THW Cu**

70. What is the calculated load for the grounded (neutral) conductor (Standard Calculation)?

The neutral load is computed per 220, Parts I and II. In general, the maximum unbalance shall be the maximum net computed load between the neutral and any one hot conductor per 220.61. To calculate the maximum unbalance, all loads connected to the neutral shall be computed. See Informative Annex D, Examples D1(a), D1(b), D2(a), and D2(b). There is no Optional Calculation method for calculating neutral loads.

220.40, Table 220.42: General Lighting:
50′ × 42′ × 3 VA/sq ft = 6300 VA
220.52(A)(B): Small Appliance and Laundry:
1500 VA × 3 = 4500 VA
Applying Demand Factors 10,800 VA
Table 220.42: First 3000 VA
or less × 100% = 3000 VA
Next 3001 VA to 120,000 VA × 35%
10,800 VA – 3000 VA =
7800 VA × 35% = 2730 VA
General Lighting and Small Appliance Load: 3,000 VA + 2,730 VA = 5730 VA
Table 220.55; 220.61(B): Range:
8400 VA × 70% = 5880 VA
220.61(B): Dryer: 5000 VA × 70% = 3500 VA
220.53: 1200 VA Dishwasher = 1200 VA
220.82: 864 VA Food Waste Disposer = 864 VA
17,174 VA

$$Load = \frac{17{,}174\ \text{VA}}{240\ \text{V}} = 71.55 \text{ or } 72$$

Calculated Load for Grounded Conductor = **72 A**

71. What is the total load for a dwelling with a central A/C unit drawing 30 A at 240 V, 1ϕ with a 10 kW, 240 V, 1ϕ central space-heating (Optional Calculation)?

220.82(C)(1): 100% nameplate rating of A/C:
30 A × 240 V = 7200 VA

220.82(C)(4): 65% nameplate rating of heat:
10,000 VA × 65% = 6500 VA
Demand Load Without A/C and Heat = 20,146 VA
220.60: Largest Load: A/C = 7200 VA
27,346 VA

$$Load = \frac{27{,}346\ \text{VA}}{240\ \text{V}} = 114\ \text{A}$$

Total Load = **114 A**

72. What is the total load for a dwelling with a central 40 A, 240 V, 1ϕ AC and 15 kW, 240 V, 1ϕ central electric space heating (Optional Calculation)?

220.82(C)(1): 100% nameplate rating of A/C:
40 A × 240 V = 9600 VA

220.82(C)(4): 65% nameplate rating of heat:
15,000 VA × 65% = 9750 VA

Demand Load Without A/C and Heat =
20,146 VA

220.60: Largest Load: Heat = 9750 VA

29,896 VA

$$Load = \frac{29{,}896 \text{ VA}}{240 \text{ V}} = 125 \text{ A}$$

Total Load = **125 A**

73. What is the total load for a dwelling with no air conditioning and three separately controlled 2 kW, 240 V, 1ϕ electric space heating units (Optional Calculation)?

220.82(C)(4): 65% nameplate rating of heat:
3 × 2000 VA × 65% = 3900 VA

Demand Load Without A/C and Heat =

$$Load = \frac{24{,}046 \text{ VA}}{240 \text{ V}} = 100 \text{ A}$$

Total Load = **100 A**

74. What is the total load for a dwelling with no air conditioning and five 1.5 kW, 240 V, 1ϕ separately controlled space heating units (Optional Calculation)?

220.82(C)(5): 40% nameplate rating of heat:
5 × 1500 VA × 40% = 3000 VA

Demand Load Without A/C and Heat =

$$Load = \frac{23{,}416 \text{ VA}}{240 \text{ V}} = 96 \text{ A}$$

Total Load = **96 A**

SERVICE LOAD CALCULATIONS (STANDARD CALCULATION–220, PARTS A AND B)

The Standard Calculation for calculating the service entrance loads for a dwelling unit is more difficult to use than the Optional Calculation. Service entrance loads computed by the Standard Calculation generally require a larger service than service entrance loads calculated by the Optional Calculation.

Examples: Standard Calculation

Use the following for Problems 75 and 76:

A dwelling unit has a floor area of 2,700 sq ft with a 12 kW range, 1.2 kW dishwasher, 4.5 kW hot water heater, 4.8 kW dryer, 864 W garbage disposal, 1.5 kW garage door opener, and a 40 A central A/C unit. The service is 120/240 V, 1ϕ.

75. What is the general lighting load (Standard Calculation)?

Table 220.12: General Lighting:
2,700 VA × 3 VA/sq ft = 8100 VA

220.52(A): Small Appliance:
1500 VA × 2 = 3000 VA

220.52(B): Laundry: 1500 VA = 1500 VA

12,600 VA

Applying Demand Factors

Table 220.42: First 3000 VA or less × 100% = 3000 VA

3001 VA to 120,000 VA × 35% =
9600 VA × 35% = 3360 VA

6360 VA

General Lighting Load = **6360 VA**

76. What is the calculated load for the ungrounded (hot) conductors (Standard Calculation)?

220.12: General Lighting Load = 6360 VA
Table 220.55, Col C: Range = 8000 VA
220.53: Dishwasher: 1.2 kVA × 75%= 900 VA
Hot Water Heater: 4.5 kVA × 75% = 3375 VA
Garbage Disposal: 864 VA × 75% = 648 VA
Garage Door Opener: 1.5 kVA × 75% = 1125 VA
Air Conditioning: 40 A × 240 V = 9600 VA
220.54: Dryer: 5000 VA = 5000 VA

35,008 VA

$$Load = \frac{35{,}008 \text{ VA}}{240 \text{ V}} = 146 \text{ A}$$

Calculated Load for Ungrounded Conductors = **146 A**

CHAPTER 3

NEC® Examples

Review Questions 3-1

Name ______________________ **Date** ______________

The answers for all review questions are listed in the answer key located in the back of this book.

______ **1.** In designing circuits, the current-carrying capacity of conductors should be corrected for heat at room temperatures above ___°F.

A. 30 C. 72
B. 86 D. 90

______ **2.** The most common type of CB is ___.

A. instantaneous C. impedance type
B. inverse time D. power factor type

______ **3.** In general, ampacity for single motors is based on the ___.

A. nameplate rating C. armature amps
B. locked-rotor amps D. none of the above

______ **4.** When determining the load on the VA/sq ft basis, the floor area shall be computed from the ___ dimensions of the building.

A. outside C. all of the above
B. inside D. none of the above

______ **5.** The fill capacity of a nipple not exceeding 24″ in length is ___%.

A. 100 C. 70
B. 40 D. 60

______ **6.** ___ 12 AWG XF conductors can be installed in a ½″ EMT conduit.

A. Four C. Six
B. Seven D. Eight

______ **7.** ___ 12 AWG RHW conductors without outer covering can be installed in a 1″ RMC raceway.

A. Thirteen C. Eight
B. Nine D. Seven

______ **8.** The derating percentage for eleven current-carrying conductors is ___%.

A. 50 C. 80
B. 0 D. 70

______ **9.** The cross-sectional area of 1½″ liquidtight flexible metal conduit is ___ sq in.

A. 1.981 C. 1.610
B. 1.588 D. 2.017

______________ **10.** ___ is ambient temperature.
A. Temperature surrounding installations
B. Room temperature
C. Outside temperature
D. all of the above

______________ **11.** A bare 4 AWG Cu conductor may be concrete-encased and serve as the GEC when at least ___′ in length.
A. 25
B. 15
C. 10
D. 20

______________ **12.** The area of a 12 AWG RHW conductor without outer covering is ___ sq in.
A. 0.0135
B. 0.0206
C. 0.0230
D. 0.0260

______________ **13.** The diameter of a 10 AWG solid aluminum conductor is ___″.
A. 0.011
B. 0.008
C. 0.102
D. none of the above

______________ **14.** Generally, branch-circuit conductors supplying a single motor shall have an ampacity not less than ___% of the motor FLC rating.
A. 110
B. 115
C. 125
D. 200

______________ **15.** A ___ THWN service-entrance conductor is permitted for a 100 A, 120/240 V, 1ϕ service for a dwelling.
A. 3 AWG Cu
B. 1 AWG Al
C. 4 AWG Cu
D. 1 AWG Cu

______________ **16.** Ten general-purpose receptacles added to a dwelling increases the service load by ___ VA.
A. 0
B. 1620
C. 3240
D. none of the above

______________ **17.** Fire protection emergency systems in assembly occupancies for not less than ___ people shall meet additional requirements.
A. 100
B. 150
C. 200
D. 1000

______________ **18.** FCC cable shall be covered with carpet squares not larger than ___ sq in.
A. 39.37
B. 24
C. 30
D. 12

______________ **19.** A single conductor cable installed in cable tray shall be larger than ___ AWG.
A. 1
B. 1/0
C. 2/0
D. 3/0

______________ **20.** An enclosure for a receptacle out of doors is installed in a flush-mounted box in a finished surface. The faceplate assembly must provide a ___ connection between the plate and the finished surface.
A. raintight
B. rainproof
C. waterproof
D. watertight

CHAPTER 3

NEC® Examples

Review Questions 3-2

Name ______________________ **Date** ______________

The answers for all review questions are listed in the answer key located in the back of this book.

______ **1.** A rating of ___ A is not a standard rating for a fuse.

A. 15 C. 75
B. 20 D. 125

______ **2.** There are ___ conductor strands in 1/0 AWG THHN Al wire.

A. 7 C. 15
B. 10 D. 19

______ **3.** The minimum branch circuit rating for an 8¾ kW household range shall be ___ A.

A. 40 C. 30
B. 35 D. 50

______ **4.** The diameter of bare 10 AWG Cu stranded conductor is ___″.

A. 0.102 C. 0.168
B. 0.116 D. 0.199

______ **5.** The FLC of a 5 HP, 208 V, 1ϕ motor is ___ A.

A. 15.2 C. 28
B. 16.72 D. 30.8

______ **6.** The short-circuit protection for the motor in Problem 5, using an ITCB, normally may not exceed ___ A.

A. 40 C. 70
B. 60 D. 80

______ **7.** If the nameplate of the motor in Problem 5 reads 28 A, ___ A is the value used to determine the overload protection.

A. 28 C. 38.5
B. 30.8 D. none of the above

______ **8.** The length of uncoated, stranded 12 AWG Cu conductor with a resistance of 0.41 Ω is ___′.

A. 207 C. 108
B. 432 D. 500

______ **9.** The maximum size ITCB allowed for A 120/240 V, 1ϕ, 3-wire feeder that powers a 3 HP, 230 V, 1ϕ motor and two 1 HP, 230 V, 1ϕ motors is ___ A.

A. 45 C. 50
B. 35 D. 60

__________ **10.** The minimum allowable ampacity for the feeder conductors in Problem 9 is ___ A.

A. 29 C. 33
B. 37 D. none of the above

__________ **11.** The ampacity of a 6 AWG RHW Al conductor in a room temperature of 47°C is ___ A.

A. 50 C. 41
B. 37.5 D. 49.2

__________ **12.** The maximum size OCPD permitted for 10 AWG TW Cu conductor in a 1″ conduit 23″ long with a total of 11 current-carrying conductors in an ambient temperature of 28°C is ___ A.

A. 15 C. 25
B. 20 D. 30

__________ **13.** The branch circuit load for a 12 kW household range is ___ kW.

A. 12 C. 8
B. 15 D. 10

__________ **14.** The ampacity for a 3 HP, 230 V, 1ϕ motor is ___ A.

A. 17 C. 18.7
B. 21 D. 25.5

__________ **15.** The total demand load for eight 14 kW, five 8 kW, and ten 12 kW household ranges is ___ kW.

A. 39.9 C. 292
B. 38 D. none of the above

__________ **16.** The ampacity of a 1 AWG, RHW, 167°F, Al conductor in a room temperature of 59°C is ___ A.

A. 100 C. 71
B. 67 D. 58

__________ **17.** An autotransformer rated less than 1000 V shall have the overcurrent protection set at not more than ___% of the full-load input current of the transformer.

A. 110 C. 125
B. 115 D. 250

__________ **18.** Overhead conductors other than those required for the amusement ride itself shall be maintained at a clearance of ___′ in any direction.

A. 10 C. 25
B. 15 D. 50

__________ **19.** The interior of a spray painting booth is a ___ location.

A. Class I, Division 1 C. Class II, Division 1
B. Class I, Division 2 D. Class III, Division 1

__________ **20.** Temporary wiring on a construction site shall be removed ___ upon completion of the project.

A. within 30 days C. within 60 days
B. within 7 days D. immediately

CHAPTER 3

NEC® Examples

Review Questions 3-3

Name ____________________ **Date** ____________

The answers for all review questions are listed in the answer key located in the back of this book.

________ **1.** The load is calculated at ___ VA for a dwelling with a 30 A, 240 V, 1ϕ heat pump and a 7.5 kW supplementary heat strip (Optional Calculation).

A. 9555 C. 7500

B. 12,075 D. none of the above

________ **2.** The load is calculated at ___ VA for a dwelling with a 45 A, 240 V, 1ϕ A/C unit and a 15 kW central heat unit (Optional Calculation).

A. 10,800 C. 15,000

B. 20,550 D. 25,800

________ **3.** The neutral load for general lighting, including required branch circuits, is ___ VA for a 1900 sq ft dwelling.

A. 5520 C. 7500

B. 3000 D. 6000

________ **4.** The ITCB for protecting a motor feeder that powers a 5 HP, 230 V, 1ϕ motor and one 3 HP, 230 V, 1ϕ motors shall not exceed ___ A.

A. 110 C. 80

B. 175 D. none of the above

________ **5.** The minimum allowable ampacity for the feeder conductors in Problem 4 is ___ A.

A. 62 C. 77

B. 52 D. none of the above

________ **6.** The minimum size XHHW Al ungrounded conductors permitted for a 225 A, 120/240 V, 1ϕ service for a dwelling is ___.

A. 4/0 AWG C. 250 kcmil

B. 3/0 AWG D. 300 kcmil

________ **7.** A minimum of ___ 15 A general lighting circuits are required for a 2200 sq ft dwelling.

A. three C. five

B. four D. six

________ **8.** A 90′ conductor rated at 1.21 Ω/kFT has a total resistance of ___ Ω.

A. 0.1089 C. 188.3

B. 108.9 D. none of the above

_______________ **9.** A 5760 W A/C load and 10 kW central heat adds ___ VA to the service of a dwelling (Optional Calculation).

A. 10,000 C. 5760
B. 6500 D. 15,760

_______________ **10.** A 4.5 kW clothes dryer adds ___ A to the neutral load of a dwelling (Standard Calculation).

A. 18.7 C. 14.6
B. 13.1 D. 20.8

_______________ **11.** A total of ___ KF-2, 10 AWG Cu conductors are permitted to be installed in a ½″ ENT conduit.

A. six C. eight
B. four D. seven

_______________ **12.** The minimum calculated amperage for a 7 kW, 240 V, 1ϕ household range is ___ A.

A. 23 C. 58
B. 29 D. none of the above

_______________ **13.** The diameter of compact 1 AWG THHN Al is ___″.

A. 0.450 C. 0.508
B. 0.415 D. 0.332

_______________ **14.** The allowable fill for a 3″ EMT conduit containing three current-carrying conductors is ___ sq in.

A. 3.91 C. 3.54
B. 2.29 D. none of the above

_______________ **15.** A rating of ___ kW shall be used to calculate the service ungrounded conductor for a dwelling with a 12 kW range (Optional Calculation).

A. 8 C. 12
B. 5.6 D. none of the above

_______________ **16.** A load of ___ kW shall be used to calculate the service grounded conductor for a dwelling with a 12 kW range.

A. 8 C. 12
B. 5.6 D. none of the above

_______________ **17.** The demand factor for services and feeders for five mobile homes shall be ___%.

A. 100 C. 33
B. 75 D. 39

_______________ **18.** Class 1 power-limited circuits shall be supplied from a source that has a rated output of not more than ___ V.

A. 50 C. 30
B. 125 D. 300

_______________ **19.** Panelboards equipped with snap switches rated ___ A or less shall have overcurrent protection not in excess of 200 A.

A. 15 C. 25
B. 20 D. 30

_______________ **20.** Formal interpretation procedures are found in the ___.

A. NFPA Regulations Governing Committee Projects C. Code Panel requirements
B. Technical Correlating Committee D. Underwriters Laboratories specifications

CHAPTER 3

NEC® Examples

Review Questions 3-4

Name ______________________ **Date** ______________

The answers for all review questions are listed in the answer key located in the back of this book.

A 2 HP, 115 V, 1ϕ, general duty motor with no code letter is used for Problems 1 through 7.

______ **1.** The FLC is ___ A.
- A. 30
- B. 17
- C. 24
- D. 18.7

______ **2.** The minimum size conductors allowed for the branch circuit is ___ AWG THWN Cu.
- A. 12
- B. 10
- C. 8
- D. 6

______ **3.** A ___ A rating is the maximum size overload normally allowed.
- A. 30
- B. 24
- C. 36.5
- D. 27.6

______ **4.** The maximum trip current, if this motor is protected by a thermal device, is ___ A.
- A. 27.6
- B. 33.6
- C. 37.4
- D. 40.8

______ **5.** The maximum ITCB normally allowed is ___ A.
- A. 35
- B. 30
- C. 50
- D. 60

______ **6.** The maximum TDF normally allowed is ___ A.
- A. 45
- B. 40
- C. 42
- D. 35

______ **7.** The maximum one-time fuse normally allowed is ___ A.
- A. 60
- B. 70
- C. 75
- D. 80

______ **8.** Six 13 kW household ranges have a demand load of ___ kW.
- A. 22
- B. 25
- C. 31.5
- D. 78

______ **9.** The minimum size of THW Cu conductors required for a 65 A noncontinuous load with 22 current-carrying conductors in a raceway is ___ AWG.
- A. 3
- B. 4
- C. 2/0
- D. 1/0

_______________ **10.** The ampacity of a 6 AWG RHW Al conductor in a raceway with an ambient temperature of 24°C is ___ A.

A. 50 | C. 47
B. 52.5 | D. 60

_______________ **11.** The resistance of 275′ of 12 AWG THWN Cu stranded conductor is ___ Ω.

A. 0.56 | C. 0.530
B. 0.5445 | D. 0.89

_______________ **12.** The resistance of a parallel set of 250 kcmil Al conductors in a 175′ installation is ___ Ω.

A. 14.8 | C. 0.0074
B. 0.1482 | D. none of the above

_______________ **13.** Conduit fill allowed for three conductors is ___%.

A. 55 | C. 40
B. 30 | D. 38

_______________ **14.** The minimum rating of a disconnecting means for a 3 HP, 230 V, 1ϕ motor is ___ A.

A. 60 | C. 30
B. 40 | D. 20

_______________ **15.** The neutral load for a 14 kW range is ___ VA.

A. 6160 | C. 14,000
B. 9800 | D. none of the above

_______________ **16.** A 6 AWG RHW Cu conductor in a room temperature of 113°F has an ampacity of ___ A.

A. 53 | C. 55
B. 65 | D. none of the above

_______________ **17.** For equipment operating in an ambient temperature of 52°C and drawing 50 A, ___ AWG THHN Cu conductors are required for a 50 A load.

A. 3 | C. 4
B. 8 | D. 6

_______________ **18.** The general lighting and small appliance and laundry load for a 2500 sq ft dwelling is ___ VA (without applying demand factors).

A. 5500 | C. 7500
B. 12,000 | D. none of the above

_______________ **19.** The branch-circuit load for a 6 kW range is ___ VA.

A. 4800 | C. 8000
B. 6000 | D. none of the above

_______________ **20.** The branch-circuit load for two 4 kW ovens and one 5 kW cooktop unit is ___ VA.

A. 8000 | C. 12,000
B. 8400 | D. 13,000

_______________ **21.** For a dwelling with a 400 A, 120/240 V, 3-wire, 1ϕ service, ___ kcmil, XHHW 75°C Al ungrounded conductors may be permitted.

A. 800 | C. 400
B. 600 | D. none of the above

_______________ **22.** A ___ AWG Cu GEC is required for the service in Problem 21 with a cold-water pipe serving as the grounding electrode.

A. 6 | C. 2
B. 4 | D. 1/0

_______________ **23.** A conductor with a maximum operating temperature of 167°F in an ambient temperature of 50°C has a correction factor of ___.

A. 0.58 | C. no correction required
B. 0.75 | D. none of the above

_______________ **24.** The maximum amount of 12 AWG THWN-2 conductors that can be installed in a 1″ rigid PVC, Schedule 40 conduit is ___.

A. 34 | C. 25
B. 15 | D. 11

_______________ **25.** The percent of derating factor for 42 current-carrying conductors is ___%.

A. 70 | C. 40
B. 50 | D. 35

_______________ **26.** The approximate area of 16 AWG TFFN is ___ sq in.

A. 0.0072 | C. 0.0109
B. 0.0075 | D. 0.0044

_______________ **27.** The size of the GEC is based on the ___ conductor.

A. largest grounded | C. equivalent grounded
B. largest ungrounded | D. high-leg

_______________ **28.** The heating load for a dwelling using electric heat is computed ___ (Standard Calculation).

A. at 100% | C. per 220.32
B. at 80% | D. at 125%

_______________ **29.** The largest trade size conduit recognized by the NEC® is ___″.

A. 4 | C. 8
B. 6 | D. none of the above

_______________ **30.** The ampacity of an 10 AWG RHW Cu conductor in free air at 30°C is ___ A.

A. 55 | C. 45
B. 40 | D. 50

Learner Resources
atplearningresources.com/quicklinks
Access Code: 533920

CHAPTER 4

Journeyman Electrician's Practice Licensing Exams (45 Minutes)

This chapter contains practice licensing exams with a 45-minute time limit. The answers for all practice licensing exams are listed in the answer key located in the back of this book. The NEC® references, with the answers, should be thoroughly studied.

PRACTICE LICENSING EXAMS

The practice licensing exams in this chapter are typical of the questions and problems found on examinations developed by the authority having jurisdiction or a private testing agency. Completing these questions and problems will aid in preparing applicants for the examination.

Note that the questions and problems of each exam found in this chapter are sample questions only. In the interest of classroom and study time, these practice licensing exams contain fewer questions than found on a typical journeyman electrician's examination. These practice licensing exams are designed as open book tests.

A typical journeyman electrician's examination is based on electrical theory, trade knowledge, NEC® questions, and calculations.

CHAPTER 4

Journeyman Electrician's Practice Licensing Exam (45 minutes)

Practice Licensing Exam 4-1

Name ______________________ **Date** ______________

The answers for all practice licensing exam questions are listed in the answer key located in the back of this book. Time – 45 minutes

______ **1.** A total of ___ 6 AWG THW Cu conductors are permitted in an 18 cu in. FS box.
A. zero
B. three
C. five
D. six

______ **2.** The stop button is connected in ___ for a motor control circuit.
A. series
B. parallel
C. series-parallel
D. jog position

______ **3.** The nominal phase to neutral voltage for a low-voltage wye system is ___ V.
A. 120
B. 480
C. 240
D. 208

______ **4.** The high-leg voltage to both neutral and ground on a 3ϕ, 4-wire, delta secondary is ___ V.
A. 208
B. 240
C. 480
D. 120

______ **5.** Show windows shall have a receptacle installed for each ___ linear feet.
A. 10
B. 6
C. 12
D. 20

______ **6.** In general, a(n) ___ AWG conductor is the smallest conductor that can be run in parallel.
A. 1/0
B. 3
C. 10
D. 8

______ **7.** One receptacle outlet shall be installed in hallways ___′ or longer in dwelling units.
A. 6
B. 10
C. 15
D. not required

______ **8.** A(n) ___ is permitted to be installed on a small appliance circuit.
A. clock outlet
B. refrigerator outlet
C. all of the above
D. none of the above

_______________ **9.** In general, the power requirements for a duplex receptacle in a dwelling ___.
A. require 180 VA
B. are included in the VA/sq ft requirements
C. require 1500 W
D. go by the point system

_______________ **10.** Show windows are computed on ___ VA per linear foot.
A. 200
B. 300
C. 1500
D. 1200

_______________ **11.** An ammeter is connected in ___ in a circuit.
A. parallel
B. shunt
C. series
D. series-parallel

_______________ **12.** ___ or larger determine if a bushing is required on a raceway.
A. Conduits 1″
B. Conductors 8 AWG
C. Conduits 1¼″
D. Conductors 4 AWG

_______________ **13.** A ___ is an electrode permitted for grounding.
A. metal underground water pipe
B. ground ring
C. plate electrode
D. all of the above

_______________ **14.** The maximum size wire that may be connected under a wire-binding screw with up-turn lugs is ___ AWG.
A. 10
B. 6
C. 8
D. 4

_______________ **15.** A luminaire that weighs more than ___ lb shall be supported independently from the box.
A. 30
B. 50
C. 25
D. 45

_______________ **16.** Motor overloads are sized by the ___ value.
A. nameplate
B. horsepower
C. motor speed
D. duty cycle

_______________ **17.** In general, the minimum size service for a dwelling is ___ A.
A. 100
B. 125
C. 150
D. 60

_______________ **18.** One horsepower equals ___ W.
A. 674
B. 746
C. 646
D. 476

_______________ **19.** The maximum voltage permitted for a pool light is ___ V.
A. 150
B. 15
C. 12
D. none of the above

_______________ **20.** A place of assembly is a building intended for the assembly of ___ or more persons.
A. 50
B. 100
C. 25
D. 75

CHAPTER 4

Journeyman Electrician's Practice Licensing Exam (45 minutes)

Practice Licensing Exam 4-2

Name ______________________ **Date** ______________

The answers for all practice licensing exam questions are listed in the answer key located in the back of this book. Time – 45 minutes

______ **1.** If a test lamp lights and burns continuously when placed in series with a capacitor and a suitable source of DC, this is a good indication that the capacitor is ___.
- A. fully charged
- B. shorted
- C. open
- D. fully discharged

______ **2.** Current is moved through a conductor by ___.
- A. electromotive force
- B. conductance
- C. ohms
- D. power

______ **3.** Trees shall not be used for ___.
- A. temporary wiring
- B. support of overhead conductor spans
- C. all of the above
- D. none of the above

______ **4.** The amperage rating required for a branch-circuit installed to supply an exterior sign shall be a minimum of ___ A.
- A. 15
- B. 25
- C. 20
- D. none of the above

______ **5.** Power is ___ voltage times current in a pure resistive circuit.
- A. double the
- B. always
- C. AC
- D. never

______ **6.** ___ lighting is a string of outdoor lights suspended between two points.
- A. Festoon
- B. Party
- C. Malibu
- D. none of the above

______ **7.** Wound-rotor and squirrel-cage motors are two types of ___ motors.
- A. universal
- B. synchronous
- C. induction
- D. single-phase

______ **8.** A(n) ___ motor can operate on AC or DC.
- A. torque
- B. induction
- C. capacitor-start
- D. universal

______ **9.** A good lubricant for pulling wire is ___.
- A. motor oil
- B. kerosene
- C. grease
- D. none of the above

_______________ **10.** Voltaic cells produce electrical energy via ___ reaction.
A. chemical
B. fission
C. all of the above
D. none of the above

_______________ **11.** Insulating safety hand grips on tools ___.
A. are not enough to prevent electrical shock
B. are not meant to prevent electrical shock
C. should be used with other insulating equipment
D. all of the above

_______________ **12.** A ___ is two dissimilar metals joined together.
A. Joule
B. Coulomb
C. thermocouple
D. phase union

_______________ **13.** A(n) ___ is used to measure the speed of an armature directly in rpm.
A. tachometer
B. megger
C. ohmmeter
D. chronometer

_______________ **14.** A(n) ___ motor has a wide speed range.
A. DC
B. AC
C. synchronous
D. induction

_______________ **15.** The ___ winding of a current transformer carries the most current.
A. primary
B. secondary
C. interwinding
D. coil

_______________ **16.** The best electrical conductor (of those listed) is ___.
A. copper
B. gold
C. silver
D. aluminum

_______________ **17.** Ambient temperature is the ___.
A. temperature of the wire
B. differential temperature
C. temperature of the area surrounding the wire
D. melting temperature of the insulation

_______________ **18.** General requirements for communications systems are addressed in Article ___.
A. 800
B. 890
C. 820
D. 840

_______________ **19.** The definition for *abandoned cable* is listed in ___.
A. 800.2
B. 830.2
C. 100
D. 840.2

_______________ **20.** The 120 V windings of a 120/240 V, 1ϕ motor are connected in ___ when connected to a 120 V supply.
A. series
B. parallel
C. series-parallel
D. none of the above

CHAPTER 4

Journeyman Electrician's Practice Licensing Exam (45 minutes)

Practice Licensing Exam 4-3

Name ______________________ **Date** ______________

The answers for all practice licensing exam questions are listed in the answer key located in the back of this book. Time – 45 minutes

______ **1.** A(n) ___ conductor has the highest temperature rating (of those listed).

A. UF C. THHN
B. RHW D. TW

______ **2.** The wiring method approved for installation in ducts which transport dust is ___.

A. rigid metal conduit C. IMC
B. EMT D. neither A, B, nor C

______ **3.** ___ is the best metal for magnets (of those listed).

A. Brass C. Zinc
B. Steel D. Tin

______ **4.** Electrical equipment secured to and in contact with an electrically grounded metal rack or structure provided for its support shall be ___.

A. secured C. bonded
B. considered grounded D. effectively attached

______ **5.** All metal parts within ___′ horizontally of the inside walls of a pool shall be bonded together.

A. 20 C. 10
B. 15 D. 5

______ **6.** Each cell shall be computed at ___ V in a stationary alkali storage battery.

A. 1.5 C. 1.2
B. 2 D. 3

______ **7.** ___ are not covered under the NEC®.

A. Marinas C. Coal mines
B. Aircraft hangers D. Utility company offices

______ **8.** The demand factor for a feeder supplying 23 recreational vehicle sites is ___%.

A. 53 C. 45
B. 43 D. 42

______ **9.** The smallest TC Cu cable permitted is ___ AWG.

A. 14 C. 3
B. 12 D. 18

______ **10.** The minimum branch circuit rating shall be ___ A for an 8¾ kW household range.
A. 40 C. 30
B. 35 D. 50

______ **11.** Overhead conductors for outdoor lighting shall not be smaller than ___ AWG Cu for spans longer than 50′.
A. 14 C. 10
B. 12 D. 8

______ **12.** A 277 V lighting fixture used for outdoor illumination on an office building shall ___.
A. be permitted C. be 12′ above grade
B. not be permitted D. be 15′ above grade

______ **13.** There shall be no more than ___ disconnect(s) for each NEC®-permitted service.
A. one C. four
B. three D. six

______ **14.** The ampacity of 12 AWG UF wire is ___ A.
A. 25 C. 15
B. 20 D. 30

______ **15.** An apparatus in which all the circuits are intrinsically safe is called a(n) ___.
A. approved apparatus C. hazardously approved apparatus
B. intrinsically safe apparatus D. none of the above

______ **16.** ___ wire is used for switchboards only.
A. TFE C. SIS
B. THWN D. SA

______ **17.** The ampacity of 8 AWG Romex is ___ A.
A. 35 C. 50
B. 40 D. 55

______ **18.** A temperature of 86°F is ___ 30°C.
A. the same as C. 32°F more than
B. 32°F less than D. none of the above

______ **19.** If the neutral of a 120/240 V, 1ϕ, 3-wire branch circuit opens, the source voltage is ___ V.
A. 120 C. 240
B. 208 D. none of the above

______ **20.** In a 120/240 V, 1ϕ, 3-wire branch circuit, if Line A draws 5 A and Line B draws 10 A, then ___ A is drawn on the neutral.
A. 0 C. 10
B. 5 D. 15

CHAPTER 4

Journeyman Electrician's Practice Licensing Exam (45 minutes)

Practice Licensing Exam 4-4

Name ______________________ **Date** ______________

The answers for all practice licensing exam questions are listed in the answer key located in the back of this book. Time – 45 minutes

______ **1.** Conduits through which moisture may contact energized parts shall be ___ at either or both ends.

A. open
B. arranged to drain
C. sealed
D. threaded

______ **2.** The minimum size conductor permitted to be installed in cablebus is ___.

A. 3 AWG
B. 1/0 AWG
C. 250 kcmil
D. 4/0 AWG

______ **3.** The ampacity of the line conductors from a generator to the first OCPD shall be not less than ___% of the nameplate current rating on the generator.

A. 110
B. 115
C. 125
D. 300

______ **4.** When a battery is being charged, ___ is given off.

A. carbon dioxide
B. hydrogen
C. helium
D. mercury

______ **5.** Voltage generated by the compression of certain crystals is ___.

A. hysteresis
B. thermionic emission
C. mercury vapor
D. piezoelectric effect

______ **6.** The effective value of AC voltage is ___.

A. 0.707 × maximum value
B. the RMS value
C. all of the above
D. none of the above

______ **7.** The high leg of a delta secondary shall be connected to ___.

A. Phase A
B. Phase B
C. Phase C
D. the neutral bar

______ **8.** Drywall surfaces that are damaged must be repaired so there will be no open spaces greater than ___″ at the edge of the box.

A. 1/8
B. 1/4
C. 3/16
D. 3/8

______ **9.** Metal wireways containing more than ___ current-carrying conductors at any cross section shall be derated.

A. 20 C. 40
B. 30 D. 35

______ **10.** Receptacles installed within ___′ from the outside of a shower shall have GFCI protection.

A. 3 C. 8
B. 6 D. 10

______ **11.** Edison-base plug fuses shall be classified at not more than ___ V and at 30 A and less.

A. 125 C. 300
B. 150 D. 130

______ **12.** Wire bending space at terminals in meter sockets is ___.

A. covered in the NEC® C. covered by public service commission
B. utility controlled D. a state standard

______ **13.** A receptacle in a garage used for a clothes washer shall be ___.

A. a twist-lock receptacle C. GFCI protected
B. a single receptacle D. none of the above

______ **14.** The fluorescent ballast with the quietest noise rating is Type ___.

A. Q C. A
B. F D. Q/S

______ **15.** A receptacle in a commercial bathroom shall be ___.

A. orange C. GFCI-protected
B. within 6′ of the sink D. a minimum wire size of 12 AWG Cu

______ **16.** The maximum distance between receptacles over a kitchen countertop is ___.

A. 3′ C. 6′
B. 12″ D. 24″

______ **17.** Floor receptacles installed in a living room shall not be counted as part of the required outlets unless located ___ the wall.

A. close to C. within 18″ of
B. within 20″ of D. within 2′ of

______ **18.** A 6′ section of track lighting in a dwelling would increase the branch load by ___.

A. 0 W C. 450 W
B. 1.5 A per ft D. 6 A

______ **19.** The ampacity of 16 AWG TF Cu conductor is ___ A.

A. 15 C. 8
B. 6 D. 4

______ **20.** Where practical, dissimilar metals in contact anywhere in the system using rigid metal conduit shall be avoided to eliminate the possibility of ___.

A. hysteresis C. inductive action
B. galvanic action D. coefficient effect

CHAPTER 4

Journeyman Electrician's Practice Licensing Exam (45 minutes)

Practice Licensing Exam 4-5

Name ______________________________ **Date** ______________

The answers for all practice licensing exam questions are listed in the answer key located in the back of this book. Time – 45 minutes

______________ **1.** The minimum clearance for overhead service conductors above a public driveway is ___′.

A. 15 C. 12
B. 18 D. 10

______________ **2.** The minimum clearance for service conductors over residential driveways not subject to truck traffic is ___′.

A. 10 C. 15
B. 12 D. 18

______________ **3.** Dry-type transformers located indoors and rated 112.5 kVA or less shall be at least ___″ from combustible material.

A. 12 C. 24
B. 18 D. 15

______________ **4.** Type NM cable must clear a scuttle hole by ___′.

A. 0 C. 6
B. 2 D. 5

______________ **5.** When a storage battery loses its charge and must be recharged, the recharging current must come from a(n) ___ source.

A. AC C. full power
B. DC D. transformer

______________ **6.** The definition for *broadband* is listed in ___.

A. 100 C. 800.2
B. 810.2 D. 840.2

______________ **7.** The largest size EMT permitted is ___″.

A. 4 C. 5
B. 6 D. 8

______________ **8.** Branch-circuit conductors supplying two outlets for fixed-resistance space heaters in a dwelling unit shall be rated not over ___ A.

A. 20 C. 50
B. 15 D. 30

__________ **9.** A ___ in a dwelling unit shall be supplied by the two 20 A small-appliance circuits.

A. kitchen | C. dining room
B. pantry | D. all of the above

__________ **10.** The minimum burial depth for PVC conduit under a 4″ concrete slab is ___″.

A. 24 | C. 4
B. 18 | D. 0

__________ **11.** ___ may be connected ahead of the service disconnects.

A. Nothing | C. Service fuses
B. Surge arresters | D. both B and C

__________ **12.** All splices, joints, and the free ends of conductors are required to be covered with an insulation ___ the conductor.

A. as thick as | C. thicker than
B. twice the size of | D. half as thick as

__________ **13.** Where an AC system operating at less than ___ V is grounded at any point, the grounded conductor must be run to each service.

A. 1000 | C. 600
B. 300 | D. 1500

__________ **14.** Raceways enclosing service-entrance conductors shall be suitable for ___ locations and arranged to drain where exposed to the weather.

A. wet | C. weatherproof
B. watertight | D. sealed

__________ **15.** Conductors that are 1/0 AWG Cu conductors in a vertical raceway shall be supported at intervals not exceeding ___′.

A. 50 | C. 100
B. 75 | D. 125

__________ **16.** In a dwelling all 15 A and 25 A branch circuits in a bedroom must be protected by a(n) ___.

A. circuit breaker | C. arc-fault circuit interrupter
B. GFIC protector | D. surge circuit protector

__________ **17.** In areas where walls are frequently washed, conduit and boxes shall be ___.

A. mounted a ¼″ from the wall | C. dipped in rustproofing material
B. mounted at least 8′ from the floor | D. PVC only

__________ **18.** A receptacle to serve a countertop in a dining room of a dwelling unit shall be located ___.

A. not more than 20″ above the unit | C. not more than 12″ above the unit
B. not more than 6″ above the unit | D. as per plans

__________ **19.** The maximum length for a cord supplying a 240 V room air conditioner is ___′.

A. 6 | C. 3
B. 10 | D. 5

__________ **20.** A ___ is permitted for connecting the GEC.

A. plumbing strap | C. all of the above
B. pipe plug | D. none of the above

CHAPTER 4

Journeyman Electrician's Practice Licensing Exam (45 minutes)

Practice Licensing Exam 4-6

Name ______________________ **Date** ______________

The answers for all practice licensing exam questions are listed in the answer key located in the back of this book.
Time – 45 minutes

______ **1.** A storage battery for emergency lighting and power shall maintain not less than 87.5% of the total voltage at the total load for a period of at least ___.
- A. 2 hours
- B. 1½ hours
- C. 1 hour
- D. 30 minutes

______ **2.** On circuits of 1000 V or less, overhead spans up to 50′ shall have Cu conductors not smaller than ___ AWG.
- A. 12
- B. 10
- C. 8
- D. 6

______ **3.** ___ may be installed in a raceway containing service entrance conductors.
- A. Grounding electrode conductors
- B. Bonding jumpers
- C. all of the above
- D. none of the above

______ **4.** The ampacity for AC arc welders is determined by ___.
- A. rating plate and duty cycle
- B. kVA and nameplate
- C. secondary current
- D. none of the above

______ **5.** The minimum size conductor required for an arc projector is ___ AWG.
- A. 10
- B. 8
- C. 6
- D. 4

______ **6.** "DANGER HIGH VOLTAGE KEEP OUT" must be posted when voltage exceeds ___ V.
- A. 277
- B. 480
- C. 600
- D. 1000

______ **7.** Overcurrent devices shall be enclosed in ___.
- A. cabinets
- B. cutout boxes
- C. all of the above
- D. none of the above

______ **8.** The main bonding jumper shall be a ___.
- A. bus
- B. screw
- C. all of the above
- D. none of the above

________________ **9.** Wooden plugs driven into holes in masonry are permitted to support boxes weighing no more than ___ lb.

A. 6 C. 50
B. 10 D. none of the above

________________ **10.** The center of the handle of CBs used as switches shall be installed no higher than ___ above the floor or working platform.

A. 6′-7″ C. 8′
B. 6′ D. 10′

________________ **11.** SE Cable can be used for branch circuits where the ungrounded conductors are ___ covered type.

A. nylon C. ethylene
B. silicone D. thermoplastic

________________ **12.** The minimum bending space for one 250 kcmil conductor in a gutter is ___″.

A. 4½ C. 6
B. 5 D. 10

________________ **13.** Service conductors that are ___ are required for a 100 A, 120/240 V, 1ϕ service for a dwelling unit.

A. 3 AWG THW Cu C. 4 AWG THW Cu
B. 2 AWG TW Al D. 4 AWG XHHW Al

________________ **14.** When service entrance conductors exceed 1100 kcmil for copper, the bonding jumper shall be at least ___% of the largest phase conductor.

A. 15 C. 12½
B. 10 D. 8

________________ **15.** Transformers used to step-up voltage for general use are classified as ___ systems.

A. separately derived C. UPS
B. emergency D. standby

________________ **16.** For a continuous-duty electric motor rated more than 1 HP with a service factor of 1.05, the overloads shall be based on ___% of the FLC.

A. 115 C. 130
B. 125 D. 140

________________ **17.** Nonmetallic extensions shall be secured in place at intervals not exceeding ___″.

A. 6 C. 10
B. 8 D. 16

________________ **18.** Service-entrance cables mounted in contact with a building shall be supported at intervals not exceeding ___″.

A. 10 C. 15
B. 24 D. 30

________________ **19.** A starting switch in a motor control circuit is wired in ___ with the holding contacts.

A. series C. series-parallel
B. parallel D. none of the above

________________ **20.** Solenoids are ___ magnets.

A. permanent C. electro
B. natural D. none of the above

CHAPTER 4

Journeyman Electrician's Practice Licensing Exam (45 minutes)

Practice Licensing Exam 4-7

Name ______________________ **Date** ______________

The answers for all practice licensing exam questions are listed in the answer key located in the back of this book. Time – 45 minutes

______ **1.** Fault current is ___.

A. locked rotor amps
B. internal battery current
C. metric amps
D. none of the above

______ **2.** If the voltage measures 208 V between phase conductors in a 3ϕ system, ___ V is the measurement to the neutral.

A. 277
B. 120
C. 208
D. 230

______ **3.** Ions, electrolyte, and amp hours pertain to ___.

A. condensers
B. ballast
C. batteries
D. transformers

______ **4.** In a 3ϕ distribution system, the phase voltages are ___° apart.

A. 120
B. 360
C. 90
D. 60

______ **5.** Two wattmeters are required to measure power on a ___ system.

A. 1ϕ
B. 3ϕ
C. complex
D. demand

______ **6.** The minimum burial depth for a 20 A, 120 V, GFCI-protected residential branch circuit using UF wire is ___″.

A. 12
B. 18
C. 24
D. 6

______ **7.** The EGC is sized by the ___.

A. conduit diameter
B. ungrounded conductors
C. overcurrent device rating
D. disconnect

______ **8.** Impedance is ___.

A. inductive reactance
B. capacitive reactance
C. total resistance in a DC circuit
D. total resistance in an AC circuit

______ **9.** A ___ is used for testing specific gravity.

A. megger
B. galvometer
C. hydrometer
D. tachometer

_______________ **10.** The thickness of insulation for SIS 8 AWG wire is ___ mil.

A. 45 C. 60

B. 50 D. 75

_______________ **11.** Hanging electrical fixtures located directly above any part of the bathtub shall be installed so that the fixture is not less than ___′ above the top of the bathtub rim.

A. 8 C. 7

B. 6 D. none of the above

_______________ **12.** Angle pull dimensional requirements apply to junction boxes only when the size of the conductor is equal to or larger than ___ AWG.

A. 0 C. 3/0

B. 4 D. 6

_______________ **13.** Metal enclosures for GECs shall be ___.

A. rigid conduit only C. electrically continuous

B. watertight D. raintight

_______________ **14.** Aluminum grounding conductors used outside shall not be installed within ___″ of earth.

A. 18 C. 12

B. 24 D. none of the above

_______________ **15.** Service conductors run above the top level of a window shall be ___.

A. 3′ above window C. 8′ above window

B. considered out-of-reach D. accessible

_______________ **16.** The minimum size GEC is ___ AWG.

A. 6 C. 8

B. 10 D. 4

_______________ **17.** ___ is not permitted to be installed above a drop ceiling used as a return air system.

A. IMC C. RMC

B. EMT D. NMC

_______________ **18.** Nail plates are required to be at least ___″ thick.

A. 1/16 C. 1/4

B. 1/8 D. 3/16

_______________ **19.** The color ___ is to be used for marking the high leg.

A. white C. gray

B. red D. orange

_______________ **20.** ___ are used to fasten an enclosure to a hollow-block wall.

A. Wood plugs C. Concrete anchors

B. Toggle bolts D. Lag bolts

CHAPTER 4

Journeyman Electrician's Practice Licensing Exam (45 minutes)

Practice Licensing Exam 4-8

Name ______________________ **Date** ____________

The answers for all practice licensing exam questions are listed in the answer key located in the back of this book.
Time – 45 minutes

________ **1.** The highest amperage at the rated voltage that an OCPD is intended to open under standard test conditions is its ___ rating.

A. ampacity
B. interrupting
C. load
D. maximum

________ **2.** A DC voltmeter may also be used to measure ___.

A. henrys
B. inductance
C. ohms
D. frequency

________ **3.** One advantage of 240 V over 120 V, based on the same wattage, is ___.

A. less voltage drop
B. more power
C. less power
D. none of the above

________ **4.** The current leads the voltage when ___.

A. reactance exceeds inductance
B. inductive reactance exceeds reactance
C. resistance exceeds capacitance
D. capacitive reactance exceeds inductive reactance

________ **5.** Heating the junction of two dissimilar alloys causes ___.

A. electrons to flow
B. arcing
C. inductance
D. fission

________ **6.** Incandescent lamps are filled with ___.

A. neon
B. hydrogen
C. nitrogen
D. dry air

________ **7.** A lamp can be controlled from three different points by ___ switches.

A. two 3-way and one 4-way
B. three 3-way
C. one DP and two 3-way
D. one SP and two DP

________ **8.** A(n) ___ is an instrument for measuring the flow of electrons.

A. ohmmeter
B. voltmeter
C. ammeter
D. wattmeter

________ **9.** A battery operates on the principle of ___ energy.

A. magnetic
B. heat
C. static
D. chemical

_______________ **10.** ___ conductors shall be used for wiring on fixture chains.

A. Solid | C. Wrapped
B. Covered | D. Stranded

_______________ **11.** Heating cables shall be identified with name or symbol, catalog number, and ___.

A. a rating in volts and watts | C. a rating in volts and amperes
B. either A or C | D. conductor color per Article 424

_______________ **12.** The largest standard ITCB is ___ A.

A. 4000 | C. 6000
B. 3000 | D. 8000

_______________ **13.** Hysteresis, eddy currents, and ampere turns are terms used when referring to ___.

A. motors | C. generators
B. lightning arresters | D. transformers

_______________ **14.** One receptacle installed on an individual branch circuit shall have a rating not less than ___% of the rating of the branch circuit.

A. 50 | C. 100
B. 80 | D. 125

_______________ **15.** A ___ A OCPD is required for a 7.5 kW, 240 V, 1ϕ strip heater.

A. 35 | C. 40
B. 60 | D. 39

_______________ **16.** The maximum resistance to ground for a single rod, pipe, or plate grounding electrode is ___ Ω.

A. infinity | C. 100
B. 25 | D. 50

_______________ **17.** Piezo electricity is caused by ___ applied to crystals.

A. chemicals | C. pressure
B. light | D. heat

_______________ **18.** The approximate diameter of 12 AWG THW is ___″.

A. 0.212 | C. 0.182
B. 0.13 | D. 0.152

_______________ **19.** Two common losses in transformer coils are eddy current and ___ losses.

A. electrolysis | C. capacitance
B. hysteresis | D. impedance

_______________ **20.** Rigid metal conduit must be buried at least ___″ in locations not otherwise specified.

A. 6 | C. 18
B. 12 | D. 24

CHAPTER 4

Journeyman Electrician's Practice Licensing Exam (45 minutes)

Practice Licensing Exam 4-9

Name ______________________ **Date** ______________

The answers for all practice licensing exam questions are listed in the answer key located in the back of this book.
Time – 45 minutes

________ **1.** ___ is the opposition to the flow of magnetic flux.

A. Henrys
B. Inductance
C. Reluctance
D. Ohms

________ **2.** A three-conductor cord for a kitchen disposal shall be not less than ___″ nor more than ___″ in length.

A. 18; 36
B. 18; 48
C. 16; 30
D. 16; 36

________ **3.** The ampacity of a capacitor's circuit conductors shall not be less than ___% of the rated amperage of the capacitor.

A. 110
B. 115
C. 125
D. 135

________ **4.** If the voltage of a circuit is constant, the amperage ___ when the resistance is increased.

A. decreases
B. remains the same
C. increases
D. drops to zero

________ **5.** Conductors supplying an outlet for a professional-type xenon projector shall not be smaller than ___ AWG.

A. 8
B. 6
C. 4
D. 3

________ **6.** Conductors of ___ AWG and larger size shall be stranded when installed in conduit.

A. 6
B. 4
C. 8
D. 10

________ **7.** A(n) ___ is used to convert AC into DC.

A. condenser
B. rectifier
C. electromagnet
D. transformer

________ **8.** The ___ is the unit of measurement for capacitance.

A. henry
B. ohm
C. farad
D. var

_______________ **9.** The ___ is the unit of measurement for capacitive reactance.

A. henry C. farad

B. ohm D. var

_______________ **10.** The maximum length for a branch circuit supplying power for an AFCI receptacle using 12 AWG conductors is ___′.

A. 50 C. 70

B. 60 D. 80

_______________ **11.** NFPA regulations governing committee projects is concerned with ___.

A. the Standards Council for the NFPA C. Code Panel selection

B. formal Code interpretation procedures D. Code Panel No. 20 only

_______________ **12.** Conductors within ___″ of a ballast compartment shall have an insulation rating not lower than 90°C.

A. 3 C. 5

B. 4 D. none of the above

_______________ **13.** A Wheatstone bridge is used to measure ___.

A. inductance C. reluctance

B. conductance D. medium and high resistance

_______________ **14.** A dynamo is ___.

A. an electromagnet C. a solid state transformer

B. used to convert mechanical energy into electrical energy D. a rectifier

_______________ **15.** If batteries are connected in series, the voltage ___.

A. increases C. is unchanged

B. decreases D. drops to zero

_______________ **16.** A fixed appliance is ___.

A. not easily moved from one place to another C. all of the above

B. fastened or otherwise secured at a specific location D. none of the above

_______________ **17.** The NEC® contains provisions considered necessary for ___.

A. efficiency C. convenience

B. future expansion D. safety

_______________ **18.** The most heat is created when AC passes through a 10 Ω ___.

A. condenser C. resistor

B. inductive coil D. capacitor

_______________ **19.** A cable tray containing service entrance conductors shall be permanently labeled SERVICE ENTRANCE CONDUCTORS every ___′ for the entire length of the cable tray.

A. 10 C. 20

B. 15 D. 25

_______________ **20.** Main bonding jumpers for AC systems are sized per Table ___.

A. 250.66 C. 250.122

B. 250.102(C)(1) D. 250.194

CHAPTER 4

Journeyman Electrician's Practice Licensing Exam (45 minutes)

Practice Licensing Exam 4-10

Name ______________________ **Date** ______________

The answers for all practice licensing exam questions are listed in the answer key located in the back of this book. Time – 45 minutes

________ **1.** A Type S fuse adapter rated at 30 A will accommodate a ___ A fuse.
A. 20 C. 35
B. 25 D. all of the above

________ **2.** True power is always volts times amps ___.
A. in an AC circuit C. where the frequency is constant
B. in a DC circuit D. none of the above

________ **3.** The general lighting load for a lodge room shall be based on ___.
A. the number of lighting units C. 1½ VA/sq ft
B. the lumen output D. 1.7 VA/sq ft

________ **4.** ___ copper is used for all covered or insulated copper conductors.
A. Soft-drawn C. Medium hard-drawn
B. Hard-drawn D. none of the above

________ **5.** ___ shall not be used on RMC for connection at couplings.
A. Running threads C. Unions
B. Aluminum fittings D. all of the above

________ **6.** Class II locations are hazardous because of the presence of ___.
A. easily ignitable fibers C. combustible dust
B. flammable gases D. flammable vapors

________ **7.** The frames of household ranges and clothes dryers are permitted to be grounded to the grounded conductor for ___.
A. existing installations C. mobile homes
B. recreational vehicles D. all of the above

________ **8.** Volume deductions for combinations of conductor sizes in a box shall be based on ___.
A. insulation type C. Table 314.16(A)
B. Table 314.16(B) D. Ch. 9, Table 4

_______________ **9.** The phase-to-neutral voltage of a 3ϕ, 4-wire, 480 V, wye-connected secondary is ___ V.

A. 208 C. 240
B. 120 D. 277

_______________ **10.** Three-way and four-way switches shall be wired so that all switching occurs only in the ___ circuit conductor.

A. ungrounded C. neutral
B. grounded D. none of the above

_______________ **11.** An autotransformer has ___.

A. one winding C. three windings
B. two windings D. a rotor

_______________ **12.** AC/DC general-use snap switches are allowed for inductive loads not exceeding ___% of the current rating of the switch at the applied voltage.

A. 50 C. 125
B. 80 D. 200

_______________ **13.** The minimum cover for PVC under a building is ___″.

A. 6 C. 12
B. 18 D. 0

_______________ **14.** The unit of measurement for inductance is the ___.

A. ohm C. henry
B. mho D. farad

_______________ **15.** A ___ is used to measure specific electrical quantities.

A. galvanometer C. multimeter
B. hydrometer D. VM

_______________ **16.** The ampacity of an 8 AWG bare conductor installed with two insulated 8 AWG THW Cu conductors in the same raceway is ___ A.

A. 40 C. 50
B. 70 D. 55

_______________ **17.** Transformer output is rated in ___.

A. henrys C. kilovolt amps
B. ohms D. kilowatts

_______________ **18.** The combination of two dissimilar metals is a ___.

A. battery action C. piezo effect
B. thermocouple D. photo cell action

_______________ **19.** The EGC of conductors run in parallel shall be sized per ___.

A. 250.94 C. Table 310.16
B. 250.122 D. 310(H)

_______________ **20.** A branch-circuit supplying a 120 gal. fixed storage water heater shall have a rating not less than ___% of the nameplate rating.

A. 80 C. 125
B. 50 D. 150

CHAPTER 4

Journeyman Electrician's Practice Licensing Exam (45 minutes)

Practice Licensing Exam 4-11

Name ______________________ **Date** ______________

The answers for all practice licensing exam questions are listed in the answer key located in the back of this book. Time – 45 minutes

________ **1.** EMT couplings and connectors used in masonry shall be ___ type.

A. concretetight C. weatherproof
B. watertight D. die cast metal

________ **2.** Where ___ cable terminates, an approved seal shall be provided immediately after stripping to prevent the entrance of moisture into the insulation.

A. TC C. SE
B. MI D. MC

________ **3.** A(n) ___ is a protective device for limiting surge voltages by discharging surge amperage.

A. GFCI CB C. current-limiting fuse
B. ITCB D. surge arrester

________ **4.** The allowable ampacity for each conductor of six current-carrying conductors shall be reduced to ___%.

A. 80 C. 70
B. 160 D. 125

________ **5.** ___ 12 AWG conductors total are allowed in a 4″ × 1¼″ square box.

A. Ten C. Eight
B. Nine D. Twelve

________ **6.** A(n) ___ is not required to be grounded.

A. electrolytic cell C. portable hand lamp
B. aquarium D. snow blower

________ **7.** Snap switches mounted to a box by ___ screws are not permitted.

A. drywall C. sheet metal
B. wood D. all of the above

________ **8.** The maximum number of 12 AWG TFN conductors permitted for a ⅜″ flexible metal conduit with outside fittings is ___.

A. three C. five
B. four D. six

__________ **9.** Multiwire branch-circuits shall supply only ___ loads.
A. line-to-neutral
B. line-to-phase
C. line-to-ground
D. phase-to-phase

__________ **10.** ___ electrons are electrons that are easily moved.
A. Orbiting
B. Ring
C. Free
D. Loose

__________ **11.** Portions of an interior raceway system exposed to widely different temperatures shall be ___.
A. coated
B. isolated
C. sealed with an approved material
D. corrected for temperature

__________ **12.** Electromotive force can be raised or lowered by using a ___.
A. rectifier
B. transformer
C. capacitor
D. magnet

__________ **13.** For other than a totally enclosed switchboard or switchgear, a space of not less than ___′ from a combustible ceiling shall be maintained.
A. 6
B. 8
C. 3
D. 4

__________ **14.** Informative Annex J.6.1 in the NEC® refers to ___.
A. J-boxes
B. judicial buildings
C. AHJ regulations
D. ADA regulations on forward reach

__________ **15.** Short radius elbows containing ___ AWG conductors or smaller shall not contain splices.
A. 3
B. 1/0
C. 6
D. 4

__________ **16.** Each ___′, or fraction thereof, of a fixed multioutlet assembly used in lighting a store showroom shall be considered as an outlet of not less than 180 VA.
A. 5
B. 1
C. 2
D. 4

__________ **17.** Conductors are considered outside of a building where covered by ___″ of concrete beneath the building.
A. 6
B. 4
C. 3
D. 2

__________ **18.** Junction boxes shall be ___.
A. accessible
B. readily accessible
C. exposed
D. all of the above

__________ **19.** The general lighting load in a dwelling is based on ___.
A. a point system
B. amps per outlet
C. 180 VA
D. VA per sq ft

__________ **20.** ___ shall not be connected to the supply side of a service disconnect.
A. Surge protection
B. Meters
C. Cable limiters
D. Secondary ties

CHAPTER 4

Journeyman Electrician's Practice Licensing Exam (45 minutes)

Practice Licensing Exam 4-12

Name ______________________ **Date** ______________

The answers for all practice licensing exam questions are listed in the answer key located in the back of this book. Time – 45 minutes

________ **1.** Two or more ground rods shall be not less than ___′ apart.

A. 2 | C. 5
B. 4 | D. 6

________ **2.** In a vertical raceway, 2/0 AWG Al conductors shall be supported at intervals not exceeding ___′.

A. 100 | C. 180
B. 200 | D. 80

________ **3.** PVC shall not be installed in an ambient temperature exceeding ___°F.

A. 122 | C. 140
B. 167 | D. 190

________ **4.** The allowable percentage of tubing fill for two RHW Cu conductors is ___%.

A. 100 | C. 80
B. 40 | D. 31

________ **5.** Supplementary overcurrent protection shall ___.

A. be a substitute for branch circuit OCPD
B. be sized smaller than branch circuit OCPD
C. be readily accessible
D. not be used as a substitute for branch circuit OCPD

________ **6.** A fluorescent fixture installed in a clothes closet shall have a minimum of ___″ clearance between the fixture and the nearest point of storage.

A. 12 | C. 18
B. 6 | D. 24

________ **7.** The length of the cord for a cord-and-plug connected trash compactor shall be ___′ to ___′ from the plug to the rear of the appliance.

A. 3 ; 4 | C. 5 ; 6
B. 4 ; 6.5 | D. 6 ; 8

________ **8.** An A/C unit has a load of 7200 VA. The central electric heat is rated at 15 kVA. A total of ___ VA is added for an optional service load calculation.

A. 15,000 | C. 14,430
B. 9750 | D. none of the above

______ **9.** In general, the maximum overcurrent protection for 6 AWG THWN Cu is ___ A with terminals rated for 75°C.

A. 60 C. 70
B. 65 D. 66

______ **10.** The bonding conductor for a swimming pool shall not be smaller than 8 AWG solid Cu, which is ___.

A. insulated C. bare
B. covered D. all of the above

______ **11.** Four light bulbs in a single fixture should be ___-connected.

A. series C. parallel
B. wye D. series-parallel

______ **12.** To reverse the rotation of a 3ϕ motor, ___.

A. reverse all leads C. reverse armature leads
B. reverse capacitor leads D. reverse any two line leads

______ **13.** Each commercial receptacle shall be counted as ___ VA.

A. 120 C. 180
B. 300 D. 1500

______ **14.** The demand factor for three electric ranges in a restaurant is ___%.

A. 60 C. 80
B. 70 D. 90

______ **15.** Flexible cord shall not be used as a substitute for ___ wiring.

A. temporary C. concealed
B. fixed D. either B or C

______ **16.** An alternation is ___.

A. 360° C. one hertz
B. one cycle D. ½ cycle

______ **17.** Specifications for plans ___.

A. set minimum standards C. set job schedules
B. provide work divisions D. are part of the contract

______ **18.** The number 32 in a 6-32 machine screw refers to the ___.

A. screw size C. threads per inch
B. screw pitch D. screw diameter

______ **19.** Smaller gauges of wire are pencil stripped to prevent ___.

A. cutting the wire C. nicks in the wire
B. damage to insulation D. all of the above

______ **20.** The vector sum of the phase currents is equal to ___ in a balanced, resistive 3ϕ system.

A. zero C. 1.73 × Phase A
B. Phase A × PF D. 3 × Phase A

CHAPTER 4

Journeyman Electrician's Practice Licensing Exam (45 minutes)

Practice Licensing Exam 4-13

Name ______________________ **Date** ______________

The answers for all practice licensing exam questions are listed in the answer key located in the back of this book.
Time – 45 minutes

______ **1.** A building's service ungrounded conductors are 500 kcmil Cu and its GEC is a metal water pipe. A ___ Cu GEC is required.

A. 3/0 AWG
B. 2 AWG
C. 500 kcmil
D. 1/0 AWG

______ **2.** An AC transformer arc welder has a nameplate primary current of 90 A at 240 V with a 30% duty cycle. The supply conductors shall be ___ A.

A. 50
B. 90
C. 117
D. 105

______ **3.** A device box contains two 14 AWG conductors, three 12 AWG conductors, one EGC, one duplex receptacle, and two cable clamps. A(n) ___ cu in. device box is required.

A. 21
B. 19.75
C. 18
D. 25.5

______ **4.** A device box contains four 12 AWG conductors and two EGCs. A fixture to be mounted to the box has four 16 AWG conductors. A(n) ___ cu in. device box is required.

A. 18.25
B. 15.5
C. 18
D. 11.25

______ **5.** Five 10 AWG THWN Cu current-carrying conductors are installed in a raceway with an ambient temperature of 32°C. The 10 AWG conductors shall be ___ A.

A. 24
B. 22.5
C. 26.3
D. 28

______ **6.** A 120/240 V, 1ϕ, 3-wire system has two 1.5 kVA, 120 V, 1ϕ resistive loads, one 2 HP, 230 V, 1ϕ motor, and three 1 HP, 120 V, 1ϕ motors. The current on the neutral is ___ A. (All loads are balanced as closely as possible.)

A. 16
B. 12.5
C. 32
D. none of the above

______ **7.** A ⅜″ flexible metal conduit with two ungrounded 12 AWG XHHW conductors and one bare 12 AWG EGC ___.

A. requires a fitting inside
B. requires a fitting outside
C. does not require a fitting
D. none of the above

______ **8.** A transformer arc welder with a primary current of 40 A and a 60% duty-cycle has a ___ A FLC.

A. 24
B. 40
C. 31
D. 50

_______________ **9.** The maximum OCPD for the arc welder in Problem 8 is ___ A.
A. 40 C. 75
B. 50 D. 80

_______________ **10.** A capacitor has a rated current of 18 A. The required ampacity for the circuit conductors is ___ A.
A. 18 C. 54
B. 22.5 D. 24.3

_______________ **11.** SJOO cord rated at 300 V is manufactured in AWG sizes ___.
A. 18–10 C. 14–6
B. 16–8 D. 16–12

_______________ **12.** In general, the branch-circuit selection current for a hermetic refrigerant motor-compressor shall be based on ___.
A. NEC® tables C. calculations as per Article 430
B. the equipment nameplate D. the compressor overload rating

_______________ **13.** Branch circuits that supply neon signs shall not be rated in excess of ___ A.
A. 40 C. 60
B. 50 D. 30

_______________ **14.** Definitions that apply regarding the NEC® rules for communications systems include ___.
A. block C. wire
B. cable D. all of the above

_______________ **15.** In a circuit, switches or circuit breakers shall not disconnect the ___ conductor.
A. ungrounded C. grounded
B. phase D. high leg

_______________ **16.** The definition for *electrical datum plane* is found in NEC® Article ___.
A. 810 C. 430
B. 555 D. 100

_______________ **17.** Heavy-duty track lighting is identified for uses exceeding ___ A.
A. 15 C. 12
B. 20 D. 16

_______________ **18.** The minimum bending radius for 2″ Type IGS cable is ___″.
A. 24 C. 16
B. 20 D. 18

_______________ **19.** At least one 125 V, 1ϕ, 15 or 20 A receptacle shall be installed within ___′ of an indoor commercial service and in the same room.
A. 25 C. 35
B. 50 D. 40

_______________ **20.** By definition, a place of assembly is a place designed for the assembly of ___ or more persons.
A. 75 C. 200
B. 100 D. 250

CHAPTER 4

Journeyman Electrician's Practice Licensing Exam (45 minutes)

Practice Licensing Exam 4-14

Name ______________________ **Date** ______________

The answers for all practice licensing exam questions are listed in the answer key located in the back of this book. Time – 45 minutes

______ **1.** The minimum size THWN Cu ungrounded conductors required for a 200 A, 120/240 V, 1ϕ commercial service is ___ AWG.

A. 1 C. 2/0
B. 1/0 D. 3/0

______ **2.** Three resistors are connected in series across a 120 V supply. The voltage reading across R_1 and R_2 is zero. A voltage reading across R_3 is detected. The circuit fault is ___.

A. R_3 is shorted C. impedance blocked on R_1 and R_2
B. R_3 is opened D. R_3 is grounded

______ **3.** A metal octagon box contains two 12 AWG and one 12 AWG EGC. A lighting fixture is to be attached to the octagon box. The fixture has two 16 AWG fixture wires and one 16 AWG EGC. The minimum size box required is ___.

A. 4″ × 1¼″ C. 4″ × 2⅛″
B. 4″ × 1½″ D. none of the above

______ **4.** Three resistors are connected in series across a 120 V supply. The voltage across R_1 is 0 V. The voltage across R_2 and R_3 is 60 V each. The circuit fault is ___.

A. R_1 is opened C. R_3 is grounded
B. R_1 is shorted D. none of the above

______ **5.** A dwelling unit has a 120/240 V, 300 A, 1ϕ service with THW Al conductors. The grounding electrode is a concrete-encased electrode. A ___ AWG Cu GEC is required.

A. 2 C. 1/0
B. 6 D. 4

______ **6.** The power factor of a 5 kW, 1ϕ load drawing 30 A connected to a 208 V supply is ___%.

A. 92 C. 80
B. 84.5 D. 82.5

______ **7.** The minimum size metal device box required for four 12 AWG ungrounded conductors, one grounded conductor, one EGC, and four 16 AWG fixture wires is ___.

A. 3″ × 2″ × 2¾″ C. 3″ × 2″ × 2¼″
B. 3″ × 2″ × 2½″ D. 3″ × 2″ × 2″

______ **8.** A straight-through pull-box has a 2″ conduit with four 4/0 AWG THW Cu conductors. The minimum length of the pull box is ___″.

A. 24 C. 36
B. 30 D. 16

______________ **9.** The minimum size box required for two 14 AWG ungrounded conductors, two 14 AWG grounded conductors, two 14 AWG EGCs, and two 16 AWG fixture wires is ___ cu in.

A. 13.5 C. 9
B. 15.5 D. none of the above

______________ **10.** The maximum unbalanced load in an apartment house is 250 A. The minimum computed load for the neutral is ___ A.

A. 250 C. 235
B. 104 D. none of the above

______________ **11.** Generally, the grounding electrode conductor (GEC) for communications systems shall be a minimum size of ___ AWG.

A. 14 C. 10
B. 12 D. 8

______________ **12.** An 8 AWG Cu conductor in a vertical raceway shall be supported every ___′.

A. 100 C. 80
B. 50 D. 25

______________ **13.** Vegetation shall not be used for the support of ___.

A. boxes C. overhead conductor spans
B. lighting fixtures D. raceways

______________ **14.** The ampacity of Type NM cable shall be based on ___.

A. the size of NMS cable C. 230.43
B. a 40°C temperature rating D. none of the above

______________ **15.** The thickness of ZFF fixture wire insulation is ___ mils.

A. 30 C. 45
B. 15 D. 14

______________ **16.** Conductors for hoistway door interlock wiring shall be Type ___ or equivalent.

A. TW C. SF
B. THHN D. XHHW

______________ **17.** Resistive heating elements shall be subdivided and shall not exceed ___ A.

A. 48 C. 55
B. 50 D. 60

______________ **18.** Live parts of a generator operating at more than ___ VAC to ground shall not be exposed to accidental contact.

A. 50 C. 300
B. 150 D. 480

______________ **19.** The nominal voltage for an alkali type battery shall be computed at ___ V per cell.

A. 1.5 C. 2.5
B. 2 D. 1.2

______________ **20.** The definition for *high voltage* is listed in Article ___.

A. 110 C. 490
B. 100 D. 700

CHAPTER 4

Journeyman Electrician's Practice Licensing Exam (45 minutes)

Practice Licensing Exam 4-15

Name ______________________ **Date** ______________

The answers for all practice licensing exam questions are listed in the answer key located in the back of this book. Time – 45 minutes

______ **1.** The minimum size masonry box required for six 12 AWG conductors, two internal cable clamps, and one SP switch is ___ cu in.

A. 12 C. 15
B. 18 D. 21

______ **2.** The area of allowable fill for a 1¼″ EMT conduit 23″ long is ___ sq in.

A. 1.38 C. 0.6
B. 1.5 D. 0.89

______ **3.** The minimum size IMC conduit for six 12 AWG THW, six 12 AWG TW, and three 10 AWG THWN conductors is ___″.

A. ¾ C. 1¼
B. 1 D. 1½

______ **4.** The calculated amperage for a 10 kW, 240 V, 1ϕ strip heater is ___ A.

A. 41.7 C. 55.6
B. 52 D. 75

______ **5.** The minimum size OCPD required for the strip heater in Problem 4 is ___ A.

A. 50 C. 70
B. 60 D. 55

______ **6.** A 120/240 V, 1ϕ service with a computed load of 550 A and a 600 A OCPD requires ___ kcmil THWN Cu conductors paralleled.

A. 350 C. 300
B. 250 D. none of the above

______ **7.** The GEC is a water pipe for the service in Problem 6. The minimum size Cu GEC required is ___ AWG.

A. 1/0 C. 2
B. 2/0 D. 3/0

______ **8.** In an apartment building with 60 units, each unit requires a 100 A, 1ϕ, 30-circuit MLO panel. The building service is a low-voltage wye distribution. The minimum size THW Cu ungrounded feeder conductors required for the unit panels are ___ AWG.

A. 4 C. 1
B. 3 D. none of the above

______ **9.** Plans show a 200′ run of 1¼″ conduit with four 6 AWG THW Cu conductors. Specifications require a pullbox every 100′ with splices in a J-box. A ___ J-box is required.

A. 4$^{11}/_{16}$″ × 2⅛″ C. 6″ × 6″ square
B. 4″ × 4″ square D. none of the above

_______________ **10.** A piece of equipment on a branch circuit draws 16 A. The supply voltage is 208 V, 1ϕ. The length of the run is 165′. The minimum size Cu conductors required is ___ AWG. (Use approximate K.)

A. 12 C. 8
B. 10 D. 6

_______________ **11.** The residual voltage of a capacitor shall be reduced to ___ V nominal or less within one minute after the capacitor is disconnected.

A. 50 C. 120
B. 80 D. 75

_______________ **12.** Conductors for circuits operating at less than 50 V shall not be smaller than ___ AWG.

A. 16 C. 14
B. 18 D. 12

_______________ **13.** A disconnecting means of adequate capacity for at least ___% of the input required for the momentary rating of X-ray equipment shall be provided in the supply circuit.

A. 50 C. 80
B. 125 D. 250

_______________ **14.** The largest conductor permitted to be installed in a cellular concrete floor raceway is ___ AWG.

A. 2/0 C. 3
B. 1/0 D. 1

_______________ **15.** A(n) ___ is a device without rotating parts that is sized for a given 3ϕ load and permits operation from a 1ϕ supply.

A. rotary-phase converter C. static-phase converter
B. MG set D. autotransformer

_______________ **16.** An equipment enclosure containing a gasoline liquid is a ___ location.

A. Class I, Division 1 C. Class II, Division 1
B. Class I, Division 2 D. Class III, Division 1

_______________ **17.** CATV stands for ___.

A. cable antenna television C. company antenna television
B. community antenna television D. cable area television

_______________ **18.** Electrical equipment installed above a lay-in ceiling shall have a minimum size opening of ___.

A. 22″ × 22″ C. 25″ × 30″
B. 20″ × 21″ D. 30″ × 30″

_______________ **19.** The high leg phase on a 3ϕ, 4-wire, delta-connected system shall be connected to the ___ phase.

A. A C. C
B. B D. N

_______________ **20.** Low-voltage suspended ceiling power distribution systems can be used for ___ locations.

A. residential C. industrial
B. commercial D. all of the above

CHAPTER 4

Journeyman Electrician's Practice Licensing Exam (45 minutes)

Practice Licensing Exam 4-16

Name ______________________ **Date** ______________

The answers for all practice licensing exam questions are listed in the answer key located in the back of this book. Time – 45 minutes

______ **1.** An installation has a 200 A switch, 150 A fuse, and 1/0 AWG THWN Cu conductors. A(n) ___ AWG Cu EGC is required.

A. 8 | C. 1/0
B. 6 | D. 4

______ **2.** A branch circuit is protected by a 60 A fuse. The 6 AWG THWN conductors are required to be increased to 4 AWG due to voltage drop. The minimum size Cu EGC required is ___ AWG.

A. 10 | C. 6
B. 8 | D. none of the above

______ **3.** A general lighting circuit is wired with 10 AWG THWN Cu rated at 75°C. The maximum size OCPD permitted is ___ A.

A. 35 | C. 30
B. 25 | D. 40

______ **4.** A 10/2 AWG Type NM cable has a bare EGC. The ampacity of the bare 10 AWG conductor is ___ A.

A. 50 | C. 30
B. 25 | D. none of the above

______ **5.** A 240 V, 1ϕ motor with a 29.4 A LRC has a maximum rating of ___ HP.

A. ½ | C. 1
B. ¾ | D. 1½

______ **6.** The AC resistance of 1000′ of 4/0 AWG Cu in PVC is ___ Ω.

A. 0.0608 | C. 0.015
B. 0.062 | D. 0.0576

______ **7.** A 1ϕ transformer with a 120 V primary and a 12 V secondary has a rated output of 0.5 kW and is 95% efficient. The primary power is ___ VA.

A. 526 | C. 475
B. 500 | D. none of the above

______ **8.** A 3-wire, 1ϕ branch circuit supplied from a 120/208 V subpanel, feeds two 15 A HID lighting circuits. The neutral shall be considered ___.

A. carrying 0 A | C. a current-carrying conductor
B. carrying 30 A | D. for derating

______________ **9.** The ampacity of a 25 HP, 230 V synchronous motor at unity power factor is ___ A.

A. 68 C. 58

B. 75 D. 53

______________ **10.** A total of ___ 15 A, 2-wire, 120 V lighting branch circuits are required to supply a show window 30′ long.

A. three C. five

B. four D. six

______________ **11.** A minimum of ___ VA shall be required for each lot in a mobile home park.

A. 16,000 C. 25,000

B. 20,000 D. 30,000

______________ **12.** The volume of a wiring enclosure (box) shall be based on the total ___.

A. box size C. volume of the assembled sections

B. box cubic inch size D. number of conductors

______________ **13.** Type MI cable shall be supported at intervals not exceeding ___′.

A. 4 C. 6

B. 4½ D. 5

______________ **14.** A ___ is a DC device consisting of one or more 2-pole, 2-wire, nonpolarized, nongrounding-type receptacles intended to be used on DC circuits only.

A. bull switch C. 2-pole, 2-wire block

B. DC spider D. plugging box

______________ **15.** ___ construction is a system in which the concealed parts of the processes of manufacture cannot be inspected without disassembly or damage.

A. Closed C. Open

B. Concealed D. Non-accessible

______________ **16.** The OCPDs for 18 AWG stranded wire used in low-voltage circuit wiring for a park trailer shall not exceed ___ A.

A. 15 C. 10

B. 6 D. 8

______________ **17.** Coaxial cable for CATV and radio distribution systems shall be separated at least ___″ from conductors of any electric light or power circuits.

A. 4 C. 8

B. 6 D. 10

______________ **18.** The percent of fill for a raceway with one conductor is ___%.

A. 53 C. 31

B. 80 D. 40

______________ **19.** The area within ___′ horizontally from aircraft power plants or aircraft fuel tanks shall be classified as a Class I, Division 2 location.

A. 5 C. 8

B. 10 D. 20

______________ **20.** According to the NEC®, more than ___ nominal volts is considered high voltage.

A. 277 C. 500

B. 480 D. 1000

Learner Resources
atplearningresources.com/quicklinks
Access Code: 533920

CHAPTER 5

Journeyman Electrician's Practice Licensing Exams (180 Minutes)

This chapter contains practice licensing exams with a 180-minute time limit. The answers for all practice licensing exams are listed in the answer key located in the back of this book. The NEC® references, with the answers, should be thoroughly studied.

PRACTICE LICENSING EXAMINATIONS

The practice licensing examinations in this chapter are typical of the questions and problems found on examinations developed by the authority having jurisdiction (AHJ) or a private testing agency. Working through these questions and problems will aid in preparing for the examination. Note that the questions and calculation problems of the practice licensing exams in this chapter are sample questions only. Practice licensing exams are designed as open book tests.

Each practice licensing exam consists of 80 questions. Each exam contains questions based on trade knowledge, NEC® calculations, and questions from the NEC® Code book. Each practice licensing exam has a three hour time limit. This allows two minutes, fifteen seconds per question (3 hr = 180 min ÷ 80 questions = 2 min, 15 sec).

Examination Preparation

When preparing for an exam, it is important to learn as much as possible about the exam. It will help to obtain information from the certifying agency about the exam or talk with individuals who have successfully passed the exam. Also, verify that the exam is the correct exam for the certification expected.

Exam preparation should be paced. Spend time researching topics covered on the exam and develop a study schedule to focus on specific topics one at a time over an extended period of time. Review exam study materials over several days prior to taking the exam. Any review of the material the night before the exam should be limited. If unfamiliar with the exam location, drive to the site a few days before the exam date. Schedule a normal amount of sleep the night before the exam and allow ample time for traveling to the exam site.

While taking the exam, avoid being intimidated and remain calm. The necessary preparation for the exam has been completed, and now, that preparation will ensure the expected results. It is important to pace your work according to the time allotted. On the first pass through the questions, answer only the questions that can be answered with absolute certainty. If a question is skipped, be sure to skip that number on the answer sheet.

Read each question carefully and identify all possible answers. Know what is being asked before attempting to answer a question. Take time to verify that the proper answer blank has been filled in completely. Note that an absolute answer may not be provided, so the most appropriate answer available should be selected. If unsure of an answer, lightly mark the question number on the answer sheet. When finished with the other questions, return to the unanswered question and, if still unsure after checking the question again, eliminate the obvious incorrect answers. Make an educated guess using the remaining answers. Never leave an unanswered question.

When finished with the examination, review the entire exam and/or answer sheet for any missing information or answers, misplaced answers, or stray marks. The number of answers completed on the exam and/or answer sheet must match the number of questions.

CHAPTER 5

Journeyman Electrician's Practice Licensing Exams (180 Minutes)

Practice Licensing Exam 5-1

Name ______________________ **Date** ______________

The answers for all practice licensing exam questions are listed in the answer key located in the back of this book. Time – 180 minutes

______ **1.** The general lighting load for a town hall is calculated at ___ VA per square foot.

A. 1.4 C. 3
B. 2 D. 3½

______ **2.** The grounded conductor may be ___ in color.

A. green C. orange
B. gray D. black

______ **3.** A(n) ___ is not necessarily considered a wet location.

A. installation directly buried C. building under construction
B. vehicle washing area D. installation in concrete

______ **4.** The minimum height for a fluorescent luminaire in a tunnel operating at more than 277 V-to-ground, and not more than 600 V between conductors, is ___′.

A. 22 C. 20
B. 24 D. 18

______ **5.** Receptacles rated at 15 and 20 A, 125 V and installed ___ a kitchen countertop in a dwelling shall have GFCI protection.

A. within 3′ of the sink in C. to serve
B. within 6′ of the sink in D. none of the above

______ **6.** The outlet for a washing machine in a dwelling unit shall be installed within ___′ of the intended location.

A. 0 C. 7
B. 6 D. 8

______ **7.** Hallways in a dwelling unit ___′ or more in length are required to have a receptacle outlet.

A. 5 C. 10
B. 8 D. none of the above

______ **8.** A 125 V receptacle is required within ___′ of an A/C unit installed on a flat roof of a duplex dwelling.

A. 6 C. 10
B. 75 D. 25

_______________ **9.** In general, festoon lighting conductors shall not be smaller than ___ AWG.

A. 12 C. 8
B. 10 D. 16

_______________ **10.** A(n) ___ is an aid in reducing arcing in movable contacts.

A. spring C. spark cover
B. inductor D. none of the above

_______________ **11.** Luminaires installed under metal-corrugated sheet roof decking shall be no less than ___″ from the lowest surface of the roof decking to the top of the luminaire.

A. ¾ C. 1½
B. 1 D. 2

_______________ **12.** Any space of a health care facility where patients are intended to be examined or treated is called ___ space.

A. general care C. critical care
B. patient care D. support

_______________ **13.** The thickness of insulation for 12 AWG RHH is ___ mils.

A. 45 C. 20
B. 30 D. none of the above

_______________ **14.** Ceiling fans that do not exceed ___ lb are permitted to be supported by an approved outlet box.

A. 50 C. 40
B. 30 D. 70

_______________ **15.** A ⅜″ flexible fixture whip with outside fittings, containing three 12 AWG THWN conductors (black, white, and green), ___.

A. is permitted C. shall not be larger than ⅜″
B. is not permitted D. none of the above

_______________ **16.** The high leg shall be connected to ___ in panelboards.

A. Phase A C. Phase C
B. Phase B D. any phase, provided the conductor is orange

_______________ **17.** The total number of 3-pole CBs that may be installed in a panelboard is equal to ___.

A. 42 C. 20
B. 30 D. the panelboard's listing

_______________ **18.** Drywall shall be repaired if there are open spaces at the edge of the box, greater than ___″.

A. 1⁄16 C. ¼
B. ⅛ D. none of the above

_______________ **19.** A gang of three switches in a box for 277 V lighting containing each of the three phase conductors requires ___.

A. the grounded conductor to be labeled C. colors to be brown, orange, and yellow
B. each phase to be a different color D. a barrier installed between switches

__________ **20.** The minimum size EGC for a swimming pool is ___.

A. 8 AWG solid Cu
B. 8 AWG stranded Al
C. 12 AWGCu
D. 10 AWG Cu

__________ **21.** Switching devices on the property shall be located at least ___′ from the inside walls of a pool.

A. 10
B. 20
C. 15
D. 5

__________ **22.** Bathroom receptacles in a dwelling unit shall ___.

A. be supplied by one 20 A circuit
B. be GFCI protected
C. have no other outlets
D. all of the above

__________ **23.** A 50 A, 2-pole CB has a 60°C rating. The plans require all wiring to be THWN Cu. A ___ AWG conductor is required.

A. 8
B. 6
C. 4
D. none of the above

__________ **24.** A voltmeter is connected in ___ with a load to determine the voltage drop.

A. series
B. series-parallel
C. parallel
D. wye

__________ **25.** Receptacles for rooftop A/C units shall be located ___ of the unit.

A. within 25′
B. in sight
C. within 50′
D. within 75′

__________ **26.** ENT is not permitted to be used in ___.

A. poured concrete
B. damp locations
C. wet locations
D. direct earth burial

__________ **27.** For branch-circuit wiring located within a fire-rated drop ceiling, EMT shall be permitted to be attached to ceiling support wires by ___.

A. approved clips
B. steel tie wire
C. fire-rated ties
D. none of the above

__________ **28.** PVC conduit shall be ___ to remove rough edges.

A. reamed inside and trimmed outside
B. trimmed inside and outside
C. reamed
D. trimmed

__________ **29.** The grounded terminal for a receptacle may be identified by ___.

A. a metal coating white in color
B. the letter "W"
C. the word "white"
D. all of the above

__________ **30.** The heating effect of 5 A of AC compared to 5 A of DC is ___.

A. the same
B. 1.41 as great
C. 1.73 as great
D. 0.707 as great

_______________ **31.** A room containing a switchboard has a temperature of 38°C. This is a reading of ___°F.

A. 96.4 C. 100.4
B. 102.6 D. none of the above

_______________ **32.** A 240 V, 1ϕ feeder supplies a 2 HP, 3 HP, and 5 HP motor. In general, ___ A is the maximum size CB permitted for the feeder conductors.

A. 85 C. 100
B. 90 D. 110

_______________ **33.** A 1200 W, 240 V, 1ϕ load with an 85% power factor draws ___ A.

A. 6 C. 4.25
B. 5.88 D. 8

_______________ **34.** A 3 HP, 120 V, 1ϕ motor is ___% efficient.

A. 100 C. 90
B. 85 D. 55

_______________ **35.** A 300 A, 120/240 V, 1ϕ service for a dwelling requires a ___ AWG Cu GEC when connected to a concrete-encased system ground.

A. 2 C. 4
B. 1/0 D. 3

_______________ **36.** The maximum size CB permitted for short-circuit protection, if a 2 HP, 208 V, 1ϕ motor starts and carries the load, is ___ A.

A. 30 C. 40
B. 35 D. 25

_______________ **37.** The OCPD used for a 2 HP, 208 V, 1ϕ motor is not adequate for the motor to start. A ___ A TDF is permitted.

A. 30 C. 40
B. 35 D. 25

_______________ **38.** The OCPD selected for a 2 HP, 208 V, 1ϕ motor is not sufficient to allow the motor to start. A ___ A inverse time circuit breaker is permitted.

A. 35 C. 70
B. 60 D. 50

_______________ **39.** A dwelling has a floor area of 2300 sq ft plus a 150 sq ft garage. A total of ___ 15 A circuits are required for general lighting.

A. six C. four
B. five D. three

_______________ **40.** A dwelling unit has a 240 V, 1ϕ A/C heat pump that draws 35 A. The unit contains 7.5 kW of supplementary heat. These loads will add ___ VA to the service calculation.

A. 8400 C. 15,900
B. 13,275 D. 7500

_______________ **41.** A single-phase converter that supplies a variable load draws 34 A. The conductor ampacity is ___ A.

A. 39.1 C. 37.4
B. 85 D. 42.5

______ **42.** A ___ conductor is a conductor intentionally connected to earth.
A. phase C. grounding
B. grounded D. grounding equipment

______ **43.** A three-phase 30 HP, 460 V induction motor has a full-load current rating of ___ A.
A. 32 C. 40
B. 80 D. 88

______ **44.** The minimum headroom of working space for a motor control center is ___′.
A. 6 C. 7
B. 6½ D. 8

______ **45.** ___ can be connected to a small appliance circuit.
A. A garbage disposal C. Kitchen lighting
B. A range hood D. A dining room clock outlet

______ **46.** Service-entrance conductors shall have clearance of ___′ from the bottom of a window.
A. 3 C. 6
B. 5 D. 4½

______ **47.** The minimum size permitted of SJO or SJ cord supplying a water conditioner that is cord-and-plug-connected to a 20 A circuit is ___ AWG.
A. 16 C. 14
B. 18 D. 12

______ **48.** ___ is the total resistance in an AC circuit.
A. Reluctance C. Impedance
B. Mhos D. Inductance

______ **49.** ___ is not a standard ampere rating for fuses.
A. Thirty-five C. Forty
B. Fifty D. Fifty-five

______ **50.** The ampacity of 8 AWG FEPB conductors in free air, rated less than 1000 V, is ___ A.
A. 80 C. 70
B. 60 D. 50

______ **51.** The volume allowance required for a 12 AWG conductor in a box is ___ cu in.
A. 1.5 C. 2.00
B. 1.75 D. 2.25

______ **52.** A disconnecting means shall be provided on the supply side of all fuses over ___ V to ground.
A. 150 C. 50
B. 120 D. 125

______ **53.** The number of bends permitted in one run of RMC conduit is ___.
A. 4/90 C. 6/90
B. 5/90 D. 8/90

______ **54.** The ampere rating of the disconnecting means for a hermetically sealed motor-compressor shall be based on ___% of the nameplate rated-load current.

A. 110 C. 125
B. 115 D. 250

______ **55.** The length of a flexible cord supplying a swimming pool motor shall not exceed ___′.

A. 3 C. 6
B. 5 D. 4½

______ **56.** One horsepower equals ___ W.

A. 647 C. 746
B. 476 D. 674

______ **57.** Multiconductor non-power-limited fire alarm cable must be protected within ___ of the floor.

A. 18″ C. 7′
B. 6′ D. 24″

______ **58.** The unit of measurement for power is the ___.

A. volt C. amp
B. watt D. ohm

______ **59.** A grounding electrode conductor shall be connected to an effectively grounded metal water pipe within ___′ of the entrance point of a building.

A. 5 C. 10
B. 20 D. 12

______ **60.** Underground cable wiring is not permitted within ___′ horizontally of the inside wall of a swimming pool.

A. 10 C. 8
B. 20 D. 5

______ **61.** Where live parts of motors or controllers operating at over ___ V nominal to ground are exposed, a rubber insulating mat shall be provided.

A. 50 C. 110
B. 100 D. 150

______ **62.** A 3 HP, 1ϕ, 208 V motor is rated at ___ FLC.

A. 19.6 C. 17
B. 18.7 D. 34

______ **63.** The NEMA configuration for a 250 V, 20 A, 1ϕ receptacle is ___.

A. 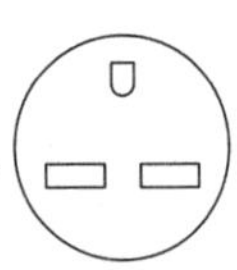 B. 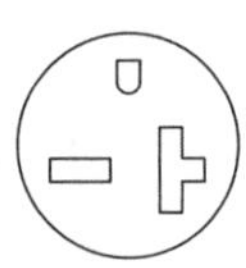 C. 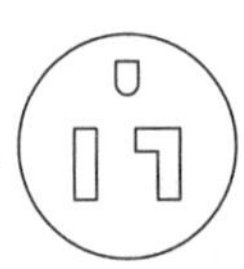 D. 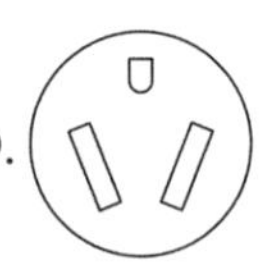

_______________ **64.** Rod and pipe electrodes shall not be less than ___′ in length.

A. 10 C. 8
B. 12 D. 6

_______________ **65.** Direct-buried conductors emerging from grade level shall be protected at least ___′ above the finished grade.

A. 3 C. 3½
B. 6 D. 8

_______________ **66.** The sheath for Type MN cable shall extend not less than ___″ inside a non-metallic box.

A. ⅜ C. ¼
B. ½ D. ⅝

_______________ **67.** In indoor wet locations where the walls are frequently washed down, raceways and boxes shall be mounted so that there is ___″ of air space between the boxes and the wall.

A. ⅜ C. ¼
B. 3⁄16 D. ⅛

_______________ **68.** A luminaire (fixture) weighing more than ___ lb shall not be supported by the screw shell of the lampholder.

A. 6 C. 8
B. 3 D. 5

_______________ **69.** The branch circuit supplying a fixed storage-type water heater that has a capacity of 120 gal. or less shall have a rating not less than ___% of the nameplate of the unit.

A. 110 C. 125
B. 115 D. 112

_______________ **70.** A non-Type IC recessed luminaire (fixture) shall be spaced ___″ from combustible material.

A. ½ C. 1
B. ¾ D. ⅞

_______________ **71.** The allowable ampacity for 14 AWG fixture wire is ___ A.

A. 20 C. 17
B. 15 D. 12

_______________ **72.** LFMC shall be fastened within ___″ of a box.

A. 8 C. 10
B. 12 D. 6

_______________ **73.** The ampacity for electric motors shall be based on ___.

A. horsepower C. NEC® tables
B. volt-amps D. service factor

_______________ **74.** Conduit seals in Class I, Division 1 locations shall be installed within ___″ of pressurize enclosures.

A. 12 C. 24
B. 18 D. 10

_______________ **75.** Each elevator car shall be supplied with ___ branch circuit(s) for lights, auxiliary lighting, and ventilation.

A. one
B. two
C. three
D. four

_______________ **76.** In general, a mobile home power supply cord shall be rated at no more than ___ A.

A. 60
B. 35
C. 45
D. 50

_______________ **77.** An outdoor motor fuel dispensing device is classified as a Class ___, Division ___ location.

A. I; 1
B. I; 2
C. II; 1
D. III; 1

_______________ **78.** In Class I, Division 1 and 2 locations, the sealing compound shall not be less than the trade size of the seal fitting and in no case less than ___″.

A. ⅝
B. ½
C. ¾
D. ⅜

_______________ **79.** The branch circuit equipment grounding conductor for a swimming pool underwater luminaire cannot be smaller than ___ AWG.

A. 10
B. 6
C. 8
D. 12

_______________ **80.** Junction boxes shall be installed so that the wiring is ___.

A. enclosed
B. covered
C. rendered accessible
D. guarded

CHAPTER 5

Journeyman Electrician's Practice Licensing Exams (180 Minutes)

Practice Licensing Exam 5-2

Name ______________________ **Date** ______________

The answers for all practice licensing exam questions are listed in the answer key located in the back of this book. Time – 180 minutes

________ **1.** The rating of any one cord-and-plug connected appliance not fastened in place must not exceed ___% of the branch circuit rating.

A. 50　　C. 80
B. 75　　D. 150

________ **2.** Flexible metal conduit used as a fixture whip to bond a fixture enclosure to a junction box shall not exceed ___′.

A. 4　　C. 6
B. 8　　D. 10

________ **3.** Rigid metal conduit 1″ in diameter shall be supported every ___′.

A. 10　　C. 14
B. 12　　D. 20

________ **4.** Liquidtight flexible metal conduit shall be used in ___″ minimum and ___″ maximum sizes.

A. ½; 4　　C. ⅜; 6
B. ½; 3　　D. ½; 6

________ **5.** The minimum length of free conductor left at each outlet or junction box shall be ___″.

A. 4　　C. 8
B. 6　　D. 10

________ **6.** The continuous current-carrying capacity of a 1″ square Cu busbar mounted in an enclosure is ___ A.

A. 500　　C. 1000
B. 750　　D. 1500

________ **7.** The maximum weight of a luminaire (light fixture) that may be mounted on the screw shell of a lampholder is ___ lb.

A. 2　　C. 10
B. 6　　D. 3

________ **8.** Premises transformers installed in modular data centers shall be of the ___ type or a type filled with a noncombustible dielectric medium.

A. wet　　C. dry
B. oil　　D. composite

_______ **9.** Instruments, pilot lights, and potential transformers shall be protected by a ___ A OCPD or less.

A. 15 | C. 30
B. 20 | D. 50

_______ **10.** Overhead Cu conductors for systems of 1000 V or less, and not over 50′ in length, shall be a minimum of ___ AWG.

A. 14 | C. 10
B. 12 | D. 8

_______ **11.** Branch-circuit conductors shall have an ampacity not less than the ___ load to be served.

A. minimum | C. peak
B. maximum | D. average

_______ **12.** The maximum number of overcurrent devices, other than mains, in a panelboard shall be ___.

A. 24 | C. based on the panelboard design
B. 36 | D. 48

_______ **13.** The liquid in a battery is called ___.

A. askarel | C. electrolyte
B. hydrogen | D. base solution

_______ **14.** A megger is used for ___.

A. reading high voltages | C. determining high resistance
B. reading coulombs | D. determining high amperes

_______ **15.** FCC cable can be installed under carpet squares no larger than ___″ square.

A. 1 | C. 1½
B. 2½ | D. 39.37

_______ **16.** The maximum rating for a Edison-based plug fuse is ___ A.

A. 20 | C. 40
B. 30 | D. 35

_______ **17.** When the number of receptacles for an office building is unknown, an additional load of ___ VA per sq ft is required.

A. 1 | C. ½
B. 2 | D. 3

_______ **18.** A terminal for a grounded conductor on a polarized plug, when not visible, shall be ___.

A. brass | C. marked with the word "white"
B. marked green | D. none of the above

_______ **19.** A reduction of ___ shall be used for a box containing one hickey and two clamps.

A. one | C. three
B. two | D. zero

______________ **20.** Raceways on the outside of a building shall be ___.

A. weatherproof and covered
B. watertight and arranged to drain
C. listed for use in wet locations and arranged to drain
D. rainproof and guarded

______________ **21.** The leads should be ___ when an ammeter is disconnected from a current transformer.

A. taped off
B. shorted
C. grounded
D. none of the above

______________ **22.** Type SFF-1 wire should be limited to use where the voltage does not exceed ___ V.

A. 500
B. 300
C. 200
D. 600

______________ **23.** Electric power production source conductors for dwellings shall have overcurrent protection located within ___′ of where the conductors are connected to the service.

A. 10
B. 12
C. 15
D. 18

______________ **24.** Where conduit is threaded in the field, the cutting die should provide ___″ taper per foot.

A. ¼
B. ½
C. ¾
D. no taper permitted

______________ **25.** Ranges and clothes dryers used in mobile homes shall be installed ___.

A. per one-family dwellings
B. with a neutral conductor
C. ungrounded
D. with a grounded and EGC

______________ **26.** ___ is not required on a motor nameplate.

A. HP
B. Manufacturer's identification
C. Watts
D. Voltage

______________ **27.** Underground conductors emerging from underground shall be in enclosures or raceways and shall be so enclosed to a point at least ___′ above finished grade.

A. 8
B. 6
C. 10
D. 15

______________ **28.** In general, busways shall be supported at intervals not exceeding ___′.

A. 3
B. 5
C. 6
D. 10

______________ **29.** In general, voltage limitation between conductors in surface metal raceway is ___ V.

A. 300
B. 600
C. 1000
D. none of the above

______________ **30.** The minimum size fixture wire is ___ AWG.

A. 16
B. 18
C. 14
D. 22

______________ **31.** A 240 V, 1ϕ branch circuit has a load of 15 A. Using 10 AWG Al conductors, ___′ is the maximum distance suggested for the run. (Use approximate K.)

A. 117.5
B. 193
C. 255
D. 167

_______________ **32.** A 5 HP, 200 V, 1φ motor has an FLC of ___ A.

A. 56 C. 30.8
B. 32.2 D. 28

_______________ **33.** A branch circuit with a load of 41 A in an ambient temperature of 53°C requires ___ AWG THWN Cu conductors.

A. 8 C. 4
B. 6 D. 3

_______________ **34.** Two balanced 120/240 V, 1φ, 3-wire branch circuits are installed in ½″ conduit. The circuit conductors are 12 AWG THW Cu. The ampacity of the ungrounded conductors is ___ A.

A. 25 C. 17
B. 15 D. 20

_______________ **35.** The branch circuit computed load for the neutral of a 15 kW household range is ___ VA.

A. 6440 C. 15,000
B. 10,500 D. 9200

_______________ **36.** A 240 V, 1φ transformer has a secondary current of 13 A. The rating of the transformer is ___ kVA. (Efficiency is 100%.)

A. 1 C. 3
B. 2 D. 4

_______________ **37.** The resistance of 150′ of 8 AWG stranded Al wire is ___ Ω.

A. 192 C. 0.189
B. 0.192 D. 19.2

_______________ **38.** A total of ___ 6 AWG THHN Cu conductors are permitted to be installed in a 1¼″ EMT conduit 22″ long.

A. 11 C. 17
B. 12 D. 9

_______________ **39.** Three loads are connected in series across a 120 V source. Each load has a resistance of 12 Ω with a current of 3.33 A. The total resistance of the circuit is ___ Ω.

A. 3 C. 6.5
B. 12 D. 36

_______________ **40.** Three resistances 6 Ω, 9 Ω, and 12 Ω are connected in parallel across a 130 V source. The total resistance is ___ Ω.

A. 2.78 C. 3
B. 27 D. none of the above

_______________ **41.** The demand load for the service calculation for a dwelling unit that has a 5 kW central heat strip and a 5 HP, 240 V, 1φ, AC compressor is ___ VA. (optional method)

A. 5000 C. 6720
B. 3250 D. 11,000

_______________ **42.** The maximum spacing between receptacles in a recreational vehicle is ___.

A. 6′ C. 12′
B. as per the furniture layout D. 18′

______ **43.** ___ are not permitted to be used to mount electrical equipment in masonry.
A. Rawl plugs
B. Wooden plugs
C. Plastic anchors
D. Steel pins

______ **44.** Switching devices shall be located at least ___′ horizontally from the inside edge of a swimming pool.
A. 5
B. 12
C. 10
D. 6

______ **45.** A ½″ × 4″ Cu busbar has ampacity of ___ A.
A. 200
B. 2000
C. 1400
D. 250

______ **46.** Circuit breakers used to switch 277 V fluorescent lighting shall be marked ___.
A. with the GFCI rating
B. with a specified NEMA number.
C. "SWD"
D. "for inductive loads only"

______ **47.** Conduit bodies in damp or wet locations shall be equipped so as to prevent ___ from entering the conduit body.
A. moisture
B. vapors
C. gases
D. fumes

______ **48.** Nonmetallic auxiliary gutters shall be supported at intervals not to exceed ___′ at each end or joint unless approved for other support intervals.
A. 4
B. 3
C. 2
D. 3½

______ **49.** A(n) ___ switch shall be permitted to be used as both motor controller and disconnecting means on a circuit whose rating does not exceed 1000 V or 100 A.
A. air-break
B. oil
C. time delay
D. auto-transformer type

______ **50.** For circuits supplying loads consisting of motor-operated utilization equipment that is fastened in place and that has a motor larger than ___ HP in combination with other loads, the total computed load shall be based on 125% of the largest motor plus the sum of the other loads.
A. ⅛
B. ⅓
C. ¼
D. ½

______ **51.** ___ means constructed so that exposure to weather such as rain will not interfere with successful operation of the apparatus under specified test conditions.
A. "Raintight"
B. "Weatherproof"
C. "Rainproof"
D. "Watertight"

______ **52.** The branch circuit load for a 15 kW 110/220 V household electric range is ___ A.
A. 68
B. 41.8
C. 62.5
D. 50

______ **53.** A 120:12 V, 1ϕ transformer that has a 500 VA secondary and is 95% efficient has ___ VA on the primary.
A. 500
B. 495
C. 526
D. 490

__________ **54.** Overhead service drop conductors may be ___ and covered.

A. insulated
B. bare
C. fluorinated ethylene propylene
D. either A or B

__________ **55.** Support for wiring located within the cavity of a fire-rated ceiling ___.

A. can be attached to the ceiling support wires
B. is allowed to the ceiling support system if tested as a fire-rated assembly
C. is not permitted
D. must have 1⁄16″ support wires

__________ **56.** Equipment mounted to a grounded metal rack ___.

A. must include a bond wire
B. shall be considered grounded
C. must be isolated from the rack
D. both A and C

__________ **57.** If a motor is connected by means of an attachment plug to a branch circuit that has no overload protection, the rating of the plug shall not exceed ___.

A. 15 A/125 V
B. 20 A/250 V
C. 20 A/125 V
D. 25 A/120 V

__________ **58.** Type USE cable may have ___.

A. a bare conductor
B. an aluminum concentric conductor applied
C. two insulated and one paper-covered conductor
D. both A and B

__________ **59.** The demand load for six 6 kW household electric dryers is ___ kW.

A. 24
B. 28.5
C. 27
D. 36

__________ **60.** Single-conductor cable used for a carnival ride shall be permitted in sizes larger than ___ AWG.

A. 2
B. 3
C. 4
D. 6

__________ **61.** The suffix 2 in RHW-2 conductors indicates ___.

A. AWG size
B. that it is permitted for continuous use at 90°C
C. mil size
D. switchboard wiring only

__________ **62.** A luminaire (light fixture) installed between 5′ and 10′ horizontally from the edge of a swimming pool shall have GFCI protection unless installed ___ above the maximum water level.

A. 5′
B. 6′-7″
C. 7′-6″
D. 7′

__________ **63.** The maximum number of disconnecting means permitted for a service is ___.

A. three
B. one
C. six
D. two

_______________ **64.** NM cable threaded through metal framing using punched or factory holes shall be ___.
A. protected by a bushing
B. protected by a grommet
C. sealed
D. both A and B

_______________ **65.** An overhead 240 V span of an open conductor shall have a minimum diagonal clearance of ___′ from a chimney.
A. 6
B. 10
C. 8
D. 3

_______________ **66.** The main bonding jumper shall be made of ___, a corrosion-resistant metal.
A. copper-clad aluminum
B. either A, C, or D
C. copper
D. aluminum

_______________ **67.** Locations protected from the weather, such as marquees and canopies, are considered ___ locations.
A. dry
B. damp
C. moist
D. wet

_______________ **68.** For final wiring purposes, a building under construction may be considered a ___ location.
A. dry
B. damp
C. moist
D. wet

_______________ **69.** An area used to wash vehicles is considered a ___ location.
A. dry
B. damp
C. moist
D. wet

_______________ **70.** "In sight from" means visible at no less than ___′.
A. 100
B. 75
C. 50
D. 25

_______________ **71.** Four hundred VA is ___ kVA.
A. 400
B. 800
C. 0.4
D. 40

_______________ **72.** At the end of a raceway, a ___ is not permitted to be used for splicing.
A. box
B. bushing
C. cabinet
D. panelboard

_______________ **73.** Voltage drop in a wire is ___.
A. wire resistance times voltage
B. a percentage of the applied voltage
C. skin effect
D. power times the voltage

_______________ **74.** The demand load for twenty-eight 8 kW household electric ranges is ___ kW.
A. 53.76
B. 22.4
C. 43
D. 56.45

_______________ **75.** The approximate K factor for copper is ___.
A. 12.9
B. 22.3
C. 13.4
D. 44.2

_______________ **76.** A circuit breaker shall be marked with the ___ in a manner that is durable and visible after installation.

A. voltage rating
B. interrupting rating
C. ampere rating
D. manufacturer name

_______________ **77.** A cord-and-plug-connected ___ is not required to be grounded.

A. dishwasher
B. hedge clippers
C. electric toaster
D. aquarium pump

_______________ **78.** RHH of size 12 AWG, without outer covering, has the same approximate area as ___.

A. RHW
B. XF
C. XFF
D. all of the above

_______________ **79.** The rating of any one piece of cord-and-plug-connected utilization equipment shall not exceed ___% of the branch-circuit rating.

A. 50
B. 80
C. 35
D. 75

_______________ **80.** The ___ conductor shall be connected to the screw shell.

A. grounded
B. ungrounded
C. grounding
D. equipment grounding

CHAPTER 5

Journeyman Electrician's Practice Licensing Exams (180 Minutes)

Practice Licensing Exam 5-3

Name ______________________ **Date** ______________

The answers for all practice licensing exam questions are listed in the answer key located in the back of this book. Time – 180 minutes

______ **1.** The allowable ampacity of 10 AWG XF Cu wire is ___ A.

A. 28 C. 25
B. 30 D. 35

______ **2.** SJ cord is not permitted to be used for ___.

A. extra-hard usage C. damp locations
B. hard usage D. pendants

______ **3.** The maximum length of a flexible cord used to supply a room A/C with a nominal 120 V rating is ___′.

A. 6 C. 10
B. 4 D. 5

______ **4.** Optical fiber cables transmit ___ for control, signaling, and communications.

A. electric impulses C. electricity
B. light D. joules

______ **5.** The minimum headroom of the working space about service equipment, switchboards, or motor control centers shall be ___′.

A. 8 C. 10
B. 8½ D. 6½

______ **6.** A bored hole in a 2 × 4 stud for a raceway should not be less than ___″ from the nearest edge.

A. ¾ C. 1
B. ⅞ D. 1¼

______ **7.** Except where computations result in a major fraction of an ampere ___ or larger, such fractions are allowed to be dropped.

A. 0.5 C. 0.7
B. 0.6 D. 0.8

_______________ **8.** Dry-type transformers installed indoors and rated 112½ kVA or less shall have a separation of at least ___″ from combustible material.

A. 6 | C. 12
B. 24 | D. 10

_______________ **9.** Fittings of the ___ type require a screwdriver for installation.

A. set-screw steel | C. set-screw die cast
B. both A and B | D. all-steel bushing

_______________ **10.** RNC is not permitted to be used for ___.

A. direct burial | C. support fixtures
B. in cinder fill | D. in ceilings

_______________ **11.** The maximum trade size permitted for rigid metal conduit is ___″.

A. 4 | C. 5
B. 6 | D. 8

_______________ **12.** The ___ is used to measure very high resistances.

A. Wheatstone bridge | C. VM
B. ohmmeter | D. megger

_______________ **13.** A voltmeter is calibrated to read the ___ value of AC.

A. effective | C. peak
B. true | D. maximum

_______________ **14.** The maximum size OCPD permitted for 12 AWG THHN Al with a load of 14 A is ___ A.

A. 15 | C. 25
B. 20 | D. 30

_______________ **15.** Electrons flow ___.

A. from positive to negative | C. counterclockwise
B. from negative to positive | D. from neutrons to protons

_______________ **16.** Two to six service disconnects shall be ___.

A. approved for damp locations | C. grouped
B. installed indoors | D. 3ϕ disconnects

_______________ **17.** Temporary wiring for general illumination lampholders shall be ___.

A. at least 10′ above finished floor | C. grounded
B. weatherproof | D. guarded

_______________ **18.** An underground feeder is installed in PVC. The circuit load is 84 A with a 100 A OCPD. The copper conductors are increased one size due to line loss. The minimum size EGC is ___ AWG.

A. 8 | C. 3
B. 6 | D. 4

_______________ **19.** A(n) ___ is not a grounding electrode system.

A. ground ring | C. underground metal gas pipe
B. metal water pipe | D. metal frame of a building

_______________ **20.** UF cable shall not be permitted to be used ___.

A. as direct burial | C. in a cable tray
B. as interior wiring | D. as service entrance cable

_______________ **21.** Fixtures shall be so constructed or installed that adjacent combustible material is not subjected to temperatures in excess of ___°C.

A. 75 | C. 185
B. 90 | D. 140

_______________ **22.** The maximum length of a bonding jumper outside a raceway or enclosure is ___′.

A. 3 | C. 25
B. 6 | D. none of the above

_______________ **23.** Conductors, splices, and taps shall not fill a metal wireway to more than ___% of its area at that point.

A. 25 | C. 125
B. 80 | D. 75

_______________ **24.** Service heads shall be listed ___.

A. for use in wet locations | C. rainproof
B. weatherproof | D. watertight

_______________ **25.** The ampacity of phase conductors from the generator terminals to the first overcurrent device shall not be less than ___% of the nameplate amperage rating of the generator.

A. 75 | C. 125
B. 115 | D. 140

_______________ **26.** The minimum size building wire permitted to be used for a 10-story building is ___ AWG.

A. 14 | C. 16
B. 12 | D. 18

_______________ **27.** The internal diameter of a 1″ rigid PVC conduit, Schedule 80 is ___″.

A. 0.936 | C. 1.255
B. 1.049 | D. 1.105

_______________ **28.** ___ shall not be used in a damp or wet location.

A. AC cable | C. EMT
B. MC cable | D. Rigid conduit

_______________ **29.** A(n) ___ is considered to be two overcurrent devices.

A. Edison-base fuse | C. lightning arrester
B. 2-pole CB | D. shunt inductor

_______________ **30.** Fire-resistive systems installed in a raceway require lubricants ___ for fire-resistive systems.

A. identified | C. listed
B. labeled | D. approved

THWN solid, uncoated Cu branch-circuit conductors sized for a 3% VD and serving a 2 HP, 230 V, 1ϕ motor are used with Problems 31 through 37.

______________ **31.** The FLC of the 2 HP motor is ___ A.

A. 12 C. 13.2
B. 13.8 D. 24

______________ **32.** The minimum size wire permitted for the motor is ___ AWG.

A. 12 C. 8
B. 10 D. 14

______________ **33.** The maximum ampacity permitted for overload protection is ___ A (based on FLC).

A. 12 C. 15
B. 13.8 D. 11.5

______________ **34.** The maximum NTDF permitted for the motor is ___ A.

A. 30 C. 35
B. 40 D. 50

______________ **35.** The maximum length of the branch-circuit conductors is ___′.

A. 95.6 C. 57.3
B. 98.6 D. 67

______________ **36.** The total resistance of the branch-circuit conductors is ___ Ω.

A. 0.2348 C. 0.4412
B. 0.75 D. 0.2935

______________ **37.** The output of the motor is ___ W.

A. 2880 C. 3600
B. 1492 D. 3168

______________ **38.** The demand load for 28, 11 kW household ranges is ___ kW.

A. 92.4 C. 43
B. 73.9 D. 48.6

______________ **39.** The branch-circuit load for a built-in 3 kW deep fryer is ___ VA.

A. 3000 C. 2400
B. 1950 D. 3750

______________ **40.** A 4 Ω resistor is connected in series to a parallel branch that contains two 8 Ω resistors connected in parallel. The total resistance is ___ Ω.

A. 12 C. 8
B. 20 D. 4.8

______________ **41.** Before approaching a person who has received an electrical shock, ___.

A. be certain the electricity is OFF C. get a first aid kit
B. get the foreman D. keep the person calm

_______________ **42.** Unless a conductor material is selected, the NEC® calculations shall be based on ___ conductors.
A. Cu-Al
B. copper-clad and Cu
C. aluminum
D. Cu

_______________ **43.** A neon tester detects ___.
A. if voltage is present
B. if a circuit is AC or DC
C. DC polarity
D. all of the above

_______________ **44.** Insulated equipment grounding conductors shall have an outer finish that is ___.
A. white
B. natural gray
C. red with a yellow stripe
D. green with a yellow stripe

_______________ **45.** A solenoid is a(n) ___.
A. electrically operated device
B. magneto
C. relay
D. thermocouple

_______________ **46.** Operating a 120 V light bulb with a voltage of 124 V results in ___.
A. shorter bulb life
B. a tripped CB
C. a blown fuse
D. dimmer light

_______________ **47.** Soldering flux is used to ___.
A. prevent corrosion
B. clean surfaces
C. melt solder
D. cool a connection

_______________ **48.** Hysteresis is ___.
A. a lagging value resulting from a change in magnetism
B. unwanted induced currents in the core of a transformer
C. impedance
D. capacitance

_______________ **49.** The machine screw with the smallest diameter is ___.
A. $^{10}/_{32} \times ^{1}/_{2}''$
B. $^{6}/_{32} \times ^{1}/_{2}''$
C. $^{8}/_{32} \times ^{1}/_{2}''$
D. $^{10}/_{24} \times ^{1}/_{2}''$

_______________ **50.** A chain wrench can be used ___.
A. for different sizes of pipe
B. in tight places
C. with one hand
D. all of the above

_______________ **51.** The best device used to recharge a battery is a(n) ___.
A. AC generator
B. DC generator
C. rectifier
D. capacitor

_______________ **52.** A covered conductor is ___.
A. encased in a material not recognized as insulation
B. enclosed in thermoplastic
C. enclosed in a nylon jacket
D. an insulated conductor

_______________ **53.** A continuous load is a load in which the maximum current is expected for ___ hr or more.

A. 1 C. 6
B. 8 D. 3

_______________ **54.** Automatic switches are actuated by an impersonal influence such as ___.

A. current C. temperature
B. pressure D. all of the above

_______________ **55.** ___ means constructed in such a way so that dust will not enter the enclosure.

A. "Dusttight" C. "NEMA 3R rating"
B. "Dustproof" D. "Sealed"

_______________ **56.** A 15 kW household electric range has a branch-circuit rating of ___ A.

A. 38 C. 50
B. 40 D. 35

_______________ **57.** A test light indicates ___.

A. voltage C. kVA
B. amperes D. continuity

_______________ **58.** ___ conductors are service conductors installed between the weather head and the service equipment.

A. Lateral C. Service entrance
B. Drop D. Service tap

_______________ **59.** Receptacles installed on a 15 A and 20 A branch circuit shall be ___.

A. ungrounded C. color coded
B. grounded D. specification grade

_______________ **60.** The least important requirement to producing a good solder connection between two stranded conductors is ___.

A. applying sufficient heat C. twisting conductors together
B. tinning each strand before soldering D. cleaning each strand

_______________ **61.** The demand load for a circuit that has a 5 kW heat strip and a 5 HP, 240 V compressor is ___ W.

A. 3250 C. 6720
B. 5000 D. 6250

_______________ **62.** In a(n) ___ circuit, the continuity of the grounded conductor shall not depend on the removal of such devices that would interrupt the continuity of the circuit.

A. branch C. multiwire branch
B. appliance D. general purpose branch

_______________ **63.** ___ is not permitted to be used for strain prevention on the joints or terminals of a cord-connected appliance.

A. Knotting the cord C. Winding the cord with tape
B. A fitting D. Nylon support string

_______________ **64.** A universal motor may not operate on ___.

A. AC C. AC or DC
B. DC D. variable frequency

_______________ **65.** Twenty-eight 6 kW household electric ranges would have a demand load of ___ kW.

A. 40.3 C. 36.96
B. 43 D. 29.5

_______________ **66.** A 350′ run supplying a 7½ HP, 1ϕ, 240 V motor fed from a 3ϕ, 240 V panel requires ___ AWG Cu wire.

A. 3 C. 4
B. 6 D. 8

_______________ **67.** A 4″ × ½″ copper busbar has an ampacity of ___ A.

A. 2000 C. 1400
B. 1500 D. 2500

_______________ **68.** The minimum cubic inch capacity of a 4″ × 1½″ square box is ___.

A. 30.3 C. 18
B. 21.5 D. 21

_______________ **69.** ___ when working on a low-voltage circuit.

A. Use insulated tools C. Always assume the circuit is hot
B. Do not work alone D. Disconnect the load

_______________ **70.** In a mobile home, ___′ is the maximum spacing between receptacles.

A. 6 C. 12
B. 8 D. 10

_______________ **71.** On a #4 drill, the number 4 indicates ___.

A. strength C. length
B. diameter D. hardness

_______________ **72.** ___ is a protection technique permitted for equipment in Class I, Division 2 locations.

A. Oil cutout C. Oil immersion
B. Oil-filled cutout D. Interrupter switch

_______________ **73.** Splices are permitted in all of the following except ___.

A. panelboards C. conduit bodies not marked with cubic inches
B. J boxes D. floor boxes

_______________ **74.** ___ is not acceptable as an equipment grounding conductor.

A. The armor of Type AC cable C. A 30 A protected conductor in flexible metal conduit
B. Intermediate metal conduit D. Cablebus framework

_______________ **75.** The overload protection for 12 AWG Cu wire is ___ A.

A. 20 C. 30
B. 25 D. 35

_______________ **76.** The maximum motor running overcurrent protection for a 7½ HP, 1ϕ, 208 V motor is ___ A.

A. 50.6
B. 44
C. 57.2
D. 55

_______________ **77.** Type S fuse adapters are designed so they ___.

A. cannot be removed
B. can be installed in parallel
C. can be removed easily
D. can be removed with a special tool

_______________ **78.** A 5 kW cooktop and a 4 kW oven have a demand load of ___ W.

A. 9000
B. 8000
C. 5850
D. 11,000

_______________ **79.** The ampacity for each of three 12 AWG XHHW Cu conductors in a wet location is ___ A.

A. 20
B. 25
C. 30
D. 35

_______________ **80.** Attachment plugs shall be constructed so that there are no exposed current-carrying parts except ___.

A. prongs
B. blades
C. pins
D. all of the above

CHAPTER 5

Journeyman Electrician's Practice Licensing Exams (180 Minutes)

Practice Licensing Exam 5-4

Name ______________________ **Date** ____________

The answers for all practice licensing exam questions are listed in the answer key located in the back of this book. Time – 180 minutes

______ **1.** A kW is ___.

A. 1000 W
B. 100 VA
C. 10 kVA
D. for rating motors

______ **2.** If you add ten duplex receptacles for general lighting in a dwelling this would increase the service load calculation by ___ kW.

A. 1.8
B. 10
C. 0
D. 1.2

______ **3.** Type MV Cable shall have ___ conductors.

A. copper
B. aluminum
C. copper-clad aluminum
D. all of the above

______ **4.** A solenoid is a(n) ___.

A. relay
B. permanent magnet
C. photo cell
D. electromagnet

______ **5.** The standard size overcurrent protection for 6 AWG TW Cu wire is ___ A.

A. 60
B. 55
C. 70
D. 75

______ **6.** Field assembled extension cord sets having ___ AWG or larger conductors shall be considered to be protected by 20 A branch-circuit protection.

A. 16
B. 18
C. A and B
D. none of the above

______ **7.** ___ is permitted to be installed in a cable tray.

A. Power and control tray cable
B. MI cable
C. PVC conduit
D. all of the above

______ **8.** The largest conductor permitted to be installed in a hollow space of a cellular metal floor is ___ AWG.

A. 1/0
B. 6
C. 8
D. 10

_______________ **9.** An AC transformer welder shall have an OCPD rated or set at not more than ___% of the rated primary current of the welder.

A. 125 C. 300
B. 80 D. 200

_______________ **10.** The maximum size Cu GEC required for a service is ___.

A. 250 kcmil C. 600 kcmil
B. 3/0 AWG D. 300 kcmil

_______________ **11.** Luminaires or lampholders shall have no live parts normally exposed to contact unless they are ___ type and located a minimum 8′ above the floor.

A. rosette C. cleat
B. recess D. none of the above

_______________ **12.** A rating of ___ A is not a standard size fuse.

A. 110 C. 75
B. 125 D. 250

_______________ **13.** Multiple grounding electrodes shall be not less than ___′ apart.

A. 6 C. 12
B. 10 D. none of the above

_______________ **14.** Conductors, ___ AWG and larger, shall be stranded when installed in raceways.

A. 8 C. 4
B. 6 D. 10

_______________ **15.** ___ is permitted to be attached to a riser conduit used to support a service drop.

A. Telephone service cable C. A and B
B. Cable TV service D. none of the above

_______________ **16.** Temporary power for holiday decorative lighting is permitted for a period not to exceed ___ days.

A. 90 C. 120
B. 60 D. 30

_______________ **17.** A counter space in a kitchen ___″ or wider requires a receptacle.

A. 12 C. 20
B. 24 D. none of the above

_______________ **18.** ___ small appliance circuit(s) is/are required for a kitchen.

A. One C. Three
B. Two D. Four

_______________ **19.** Romex cable can be used on multifamily dwellings and structures for Types ___ construction.

A. II and III C. II, III, and V
B. I, II, and III D. III, IV, and V

_______________ **20.** The NEC® is intended to be suitable for mandatory application by inspecting authorities over ___.

A. electrical installations C. A and B
B. railroad installations D. none of the above

_______________ **21.** The alternating current GEC is sized by the ___.

A. service main disconnect | C. service overcurrent size
B. service drop conductors | D. none of the above

_______________ **22.** Torque, repulsion, and induction are terms used with ___.

A. ballasts | C. motors
B. transformers | D. generators

_______________ **23.** A total of ___° of bends are permitted in a run of EMT.

A. 360 | C. 320
B. 180 | D. 240

_______________ **24.** In general, conductor size for a single motor shall have an ampacity not less than ___% of the FLC rating.

A. 110 | C. 125
B. 115 | D. 0

_______________ **25.** For stationary motors of ⅛ HP or less and 300 V or less, a ___ may serve as the disconnecting means.

A. HP-rated switch | C. CB
B. fuse | D. all of the above

_______________ **26.** A space of ___′ or more shall be provided between the top of any switchboard and any combustible ceiling.

A. 3 | C. 4
B. 6 | D. 2

_______________ **27.** The smallest wall space requiring a receptacle is ___′.

A. 2 | C. 5
B. 3 | D. 6

_______________ **28.** The total 100% area for a ½″ flexible metal conduit is ___ sq in.

A. 0.630 | C. 0.317
B. 0.632 | D. 0.622

_______________ **29.** A(n) ___ size box is required for 30.3 cubic inches of conductor fill.

A. 4 × 1¼″ sq | C. FD 2⅛″
B. 4 × 2⅛″ sq | D. 3¾ × 2 × 3½″ masonry

_______________ **30.** Shore-power receptacles shall be rated not less than ___ A.

A. 20 | C. 40
B. 30 | D. 50

_______________ **31.** An apartment building has ten 12 kW ranges and eight 14 kW ranges. The demand load is ___ kVA.

A. 36.3 | C. 33
B. 18 | D. 34.65

_______________ **32.** The demand load for sixteen 6 kW household ranges is ___ kW.

A. 26.9 | C. 96.0
B. 48.5 | D. 96.5

______________ **33.** The FLC of a 240 V, 5 HP DC motor is ___ A.
A. 15 C. 10
B. 20 D. 12.9

______________ **34.** A 1 HP, 230 V motor draws 4 A. The power consumed by internal losses is ___ VA.
A. 920 C. 746
B. 174 D. 1666

______________ **35.** The demand load for nine 10 kW household ranges is ___ kVA.
A. 25 C. 24
B. 31.5 D. 28.3

______________ **36.** A 3 Ω, 6 Ω, and 9 Ω resistor is connected in parallel. R_T is ___ Ω.
A. 1.64 C. 0.61
B. 18 D. none of the above

______________ **37.** A 3 Ω, 6 Ω, and 9 Ω resistor is connected in series. R_T is ___ Ω.
A. 1.64 C. 0.61
B. 18 D. 2.64

______________ **38.** Six 10 AWG THWN Cu current-carrying conductors are installed in a raceway. The ambient temperature is 105°F. The ampacity of the conductors is ___ A.
A. 23 C. 28.7
B. 35 D. 28

______________ **39.** A heat pump and A/C unit draws 28 A at 240 V, 1ϕ. The A/H contains a 240 V, 7.5 kW heat strip. A total of ___ VA is added to the service load calculation.
A. 6720 C. 11,595
B. 7500 D. 9243

______________ **40.** A 120 V, 1ϕ branch circuit with a 12 A load has a 116′ run. Number ___ Cu conductors are required.
A. 12 C. 8
B. 10 D. 14

______________ **41.** A nipple can be ___.
A. threaded on both ends C. not threaded
B. threaded on one end D. all of the above

______________ **42.** The best way to charge a battery is by using a(n) ___.
A. DC motor C. Wheatstone bridge
B. rectifier D. AC to DC generator

______________ **43.** ___ may be used to mount a box to a hollow block wall.
A. Toggle bolts C. Lag bolts
B. Wood plugs D. Concrete anchors

______________ **44.** When a panel is being mounted, ___ shall not be used.
A. toggle bolts C. concrete anchors
B. wood plugs D. a threaded rod

_______________ **45.** A 5 kW cooktop and a 6 kW oven have a demand load of ___ kW.

A. 9000 C. 8000

B. 5860 D. 1100

_______________ **46.** The primary current is ___ A.

A. 2.5 C. 25

B. 0.3 D. 30

_______________ **47.** Service-entrance cable used for interior wiring shall comply with ___.

A. Article 314 C. Part II of Article 334

B. not permitted D. Informative Annex C

_______________ **48.** The turns ratio of this transformer is ___.

A. 10:1 C. 10:2

B. 1:10 D. 1:12

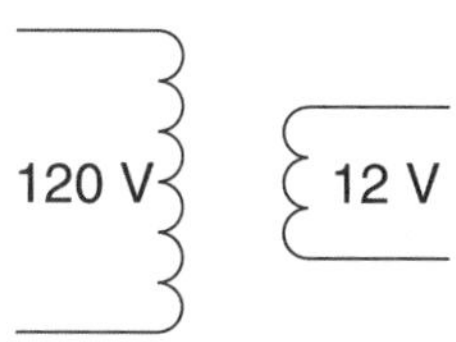

_______________ **49.** Electrical equipment secured to and in contact with a metal rack or structure provided for its support shall be considered ___.

A. secured C. bonded

B. effectively grounded D. effectively attached

_______________ **50.** Conduits through which moisture may contact energized parts shall be ___ at either or both ends.

A. open C. sealed

B. arranged to drain D. threaded

_______________ **51.** On a single-pole switch, the terminals are required to be ___.

A. not identified C. marked "SWD"

B. identified D. color coded

_______________ **52.** A Type AC cable run across the top of floor joists or within ___′ of the floor in attics shall be protected by guard strips.

A. 6 C. 10

B. 7 D. 12

_______________ **53.** Clearance for conductors not over 1000 V from signs and radio and television antennas shall not be less than ___′.

A. 3 C. 6
B. 8 D. 3½

_______________ **54.** The overload protection for 12 AWG RHH AWG Cu is ___ A.

A. 30 C. 25
B. 20 D. 35

_______________ **55.** In sight from is equipment not more than ___′ from other equipment.

A. 100 C. 50
B. 75 D. 25

_______________ **56.** Circuit breakers shall be marked for ___ rating.

A. interrupting C. voltage
B. ampere D. all of the above

_______________ **57.** An existing luminaire (light fixture) between 5′ and 10′ from the edge of a pool shall not be less than ___′ above the maximum water level.

A. 5 C. 10
B. 6 D. 8

_______________ **58.** Rigid nonmetallic conduit shall not be used ___.

A. in hazardous locations C. in theaters
B. for support of a luminaire D. all of the above

_______________ **59.** ___ shall not be used for the connection between a grounding conductor and a grounding electrode.

A. Pressure connectors C. Listed lugs
B. Exothermic welding D. Solder

_______________ **60.** The parts of a swimming pool requiring bonding shall be connected to a common bonding grid with a(n) ___ conductor not smaller that 8 AWG.

A. covered, insulated or bare C. solid
B. copper D. all of the above

_______________ **61.** Where multiple rod, pipe, and plate electrodes are used, they shall not be less than ___′ apart.

A. 6 C. 15
B. 10 D. 12

_______________ **62.** A rectifier converts ___.

A. DC to AC C. magnetism to energy
B. AC to DC D. wattage to apparent power

_______________ **63.** In a raceway containing eight current-carrying conductors, the ampacity shall be derated by ___%.

A. 80 C. 50
B. 70 D. 45

_______________ **64.** A 480 V transformer with a 240/120 V secondary has no 240 V load. The secondary has 6 kW from A–N and 6 kW from B–N. The neutral ___.

A. is the same size as the ungrounded conductor
B. is 70% smaller than the ungrounded conductor
C. has no relation to the ungrounded conductor
D. is 50% smaller than the ungrounded conductor

_______________ **65.** Two small appliance circuits, one laundry circuit, one 12 kW range, one 4 kW dryer, one 1.2 kW dishwasher, and four 1920 VA lighting loads are fed from a 3-wire, 240/120 V, 1ϕ source. Choice ___ is the best balanced load condition.

A.		B.		C.		D.	
1500	1500	1500	1920	1200	1200	1500	1920
1500	1920	1500	1920	1200	1920	1500	1920
1920	1920	1500	1920	1200	1920	1500	1920
6000	6000	6000	6000	4000	4000	1500	1920
2000	2000	4000	4000	2500	2500	1920	1200
1920	1200	1920	1200	1920	1200	4000	4000

_______________ **66.** The kVA input is mostly near ___ kVA if the transformer load is 500 VA.

A. 1500
B. 500
C. 0.5
D. 0.05

_______________ **67.** When the standard calculations in Part I and III of Article 220 are used, the A/C is calculated for the service at ___:

A. 80%
B. 125%
C. demand factor
D. full load

_______________ **68.** The NEC® allows noncontinuous metal raceways between J boxes, cabinets, and enclosures if ___.

A. EMT raceway is used for surge protection
B. rigid metal conduit is used for temporary power
C. flexible metal conduit is used for motors
D. short sections are used for cable protection

_______________ **69.** A ___ fitting is intended for termination of a cord that includes explosionproof or flameproof seals.

A. coaxial cable
B. controller connector
C. pressure connector
D. cord connector

_______________ **70.** ___ set(s) of 1 AWG THW Cu are permitted to be run in parallel.

A. One
B. Two
C. Three or more
D. Zero

_______________ **71.** To prevent ___ when working in a room full of batteries, caution should be taken not to cause sparks.

A. a fire
B. an explosion
C. contact pitting
D. an electric shock

________________ **72.** Improper makeup of splices and connections ___.

A. violates NEMA standards   C. causes heat
B. violates ANSI standards   D. increases circuit capacitance

________________ **73.** Lampholders used in branch circuits having a rating in excess of ___ A shall be of the heavy-duty type.

A. 12   C. 16
B. 15   D. 20

________________ **74.** The demand load for a 14 kW household range is ___ kW.

A. 8.8   C. 12
B. 14   D. 9.8

________________ **75.** The demand for an 8 kW household electric range is calculated at 6.4 kW. The neutral load would be ___ kW.

A. 4.48   C. 8
B. 6.4   D. 5.6

________________ **76.** The neutral current is ___ A.

A. 53.3   C. 0
B. 45   D. 8.33

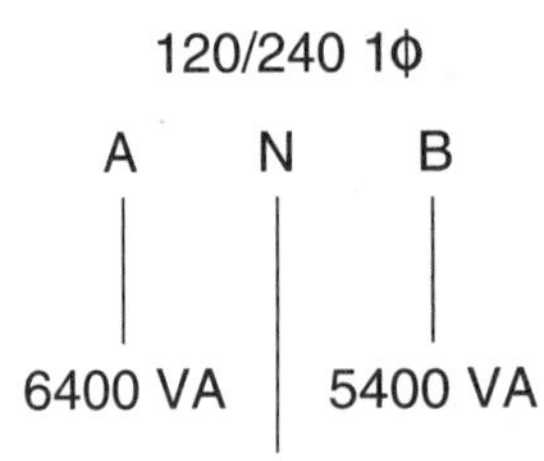

________________ **77.** The demand load for two 4.5 kVA household electric dryers is ___ kVA.

A. 10   C. 9
B. 7   D. 6.3

________________ **78.** A string of outdoor lights that is suspended between two points is ___ lighting.

A. outline   C. festoon
B. carnival   D. car lot

________________ **79.** Nichrome® wire is used in toasters over copper because ___.

A. it makes better toast   C. it has higher resistance
B. of cost   D. it has lower resistance

________________ **80.** Two 1½ HP, 115 V, 1ϕ motors require ___ AWG THW Cu conductors.

A. 12   C. 8
B. 10   D. 6

Answer Key

1 NEC® Definitions

Review Questions 1-1 ______9

1. **D**
2. **A**
3. **C**
4. **B**
5. **D**
6. **D**
7. **D**
8. **D**
9. **B**
10. **A**
11. **D**
12. **D**
13. **C**
14. **A**
15. **B**
16. **B**
17. **D**
18. **D**
19. **A**
20. **B**

Review Questions 1-2 ______11

1. **D**
2. **C**
3. **B**
4. **C**
5. **A**
6. **C**
7. **A**
8. **A**
9. **B**
10. **D**
11. **D**
12. **C**
13. **B**
14. **D**
15. **C**
16. **C**
17. **B**
18. **A**
19. **B**
20. **D**

Review Questions 1–3 _____13

1. **B**
2. **C**
3. **B**
4. **D**
5. **C**
6. **A**
7. **B**
8. **A**
9. **C**
10. **D**
11. **D**
12. **B**
13. **A**
14. **B**
15. **A**
16. **A**
17. **C**
18. **A**
19. **C**
20. **D**

2 Electrical Formulas

Review Questions 2-1 _____29

1. **A** Electrical theory
2. **A** Electrical theory
3. **B** *210.19(A), Informational Note No.3*
4. **D** Electrical theory
5. **C** Electrical theory
6. **C** $P_T = I^2 \times R$
   $P_T = (25)^2 \times 21 = 13{,}125$
   $P_T = \mathbf{13{,}125\ W}$
7. **D** $Output = HP \times 746$
   $Output = 5 \times 746 = 3730$
   $Output = \mathbf{3730\ VA}$
8. **D** $R = \dfrac{E_2}{P}$
   $R = \dfrac{240 \times 240}{10{,}000}$
   $R = \dfrac{57{,}600}{10{,}000} = 5.76$
   $R = \mathbf{5.76\ \Omega}$
9. **D** *Ch 9, Table 8:*
   1 AWG Cu = 0.154 Ω/kFT
   1 AWG Cu = 83,690 CM
   $K = \dfrac{R \times CM}{1000}$
   $K = \dfrac{0.154 \times 83{,}690}{1000}$
   $K = \dfrac{12{,}888.26}{1000} = 12.88$
   $K = \mathbf{12.88\ \Omega}$
10. **C** *Ch 9, Table 8:*
    4/0 kcmil = **211,600**
11. **D** $Cost = \dfrac{T \times W \times Cost / kWh}{1000}$
    $Cost = \dfrac{8 \times 200 \times 0.09}{1000}$
    $Cost = \dfrac{144}{1000} = 0.144$
    $Cost = \mathbf{\$0.14}$
12. **C** $R_T = R_1 + R_2 + R_3$
    $R_T = 4 + 6 + 8 = 18$
    $R_T = \mathbf{18\ \Omega}$
13. **A** $R_T = \dfrac{1}{\dfrac{1}{R_1} + \dfrac{1}{R_2} + \dfrac{1}{R_3}}$
    $R_T = \dfrac{1}{\dfrac{1}{4} + \dfrac{1}{6} + \dfrac{1}{8}}$
    $R_T = \dfrac{1}{0.250 + 0.167 + 0.125}$
    $R_T = \dfrac{1}{0.542} = 1.845$
    $R_T = \mathbf{1.845\ \Omega}$
14. **D** $HP = \dfrac{I \times E \times E_{ff}}{746}$
    $HP = \dfrac{28 \times 240 \times 0.55}{746}$
    $HP = \dfrac{3696}{746} = 4.9$
    $HP = \mathbf{5\ HP}$
15. **A** $I_P = \dfrac{kVA \times 1000}{E_P}$
    $I_P = \dfrac{1500}{120} = 12.5$
    $I_P = 12.5\ A$

Table 450.3(B), Note 1:
12.5 A × 125% = 15.6 A
Table 240.6(A):
Next higher rating = 20 A
OCPD = **20 A**

16. **D** $VD = \frac{K \times I \times M \times D}{CM}$

$VD = \frac{21.2 \times 40 \times 2 \times 210}{16{,}510}$

$VD = \frac{356{,}160}{16{,}510} = 21.57$

$VD = \mathbf{21.57\ V}$

17. **A** $R_p = \frac{R_1 \times R_2}{R_1 + R_2}$

$R_p = \frac{4 \times 3}{4 + 3}$

$R_p = \frac{12}{7} = 1.7$

$R_p = 1.7\ \Omega$
$R_T = 3\ \Omega + 1.7\ \Omega$
$R_T = \mathbf{4.7\ \Omega}$

18. **C** $R_p = \frac{1}{\frac{1}{R_1} + \frac{1}{R_2} + \frac{1}{R_3}}$

$R_p = \frac{1}{\frac{1}{2} + \frac{1}{5} + \frac{1}{10}}$

$R_p = \frac{1}{0.5 + 0.2 + 0.1}$

$R_p = \frac{1}{0.8} = 1.25$

$R_p = 1.25\ \Omega$
$R_T = 1.25\ \Omega + 6\ \Omega$
$R_T = \mathbf{7.25\ \Omega}$

19. **B** *busbar ampacity* = *t* × *w* × *constant*
busbar ampacity = 0.5 × 4 × 700
busbar ampacity = **1400 A**

20. **D** *busbar ampacity* = *t* × *w* × *constant*
busbar ampacity = 0.25 × 4 × 1000
busbar ampacity = **1000 A**

Review Questions 2-2 _____31

1. **D** $I = \frac{P}{E}$

$I = \frac{300}{130} = 2.3$

$I = \mathbf{2.3\ A}$

2. **C** $E = I \times R$
$E = 1.5 \times 40 = 60$
$E = 60\ V$

$I = \frac{E}{R}$

$I = \frac{60}{50} = 1.2$

$I = \mathbf{1.2\ A}$

3. **A** $R_T = R_1 + R_2 + R_3$
$R_T = 8 + 8 + 8 = 24$
$R_T = \mathbf{24\ \Omega}$

4. **A** $R_T = \frac{R_1}{N}$

$R_T = \frac{8}{3} = 2.66$

$R_T = \mathbf{2.66\ \Omega}$

5. **B** $R_T = \frac{R_1 \times R_2}{R_1 + R_2}$

$R_T = \frac{5 \times 10}{5 + 10}$

$R_T = \frac{50}{15} = 3.33$

$R_T = \mathbf{3.33\ \Omega}$

6. **C** $E = I \times R$
$E = 4 \times 10 = 40$
$E = \mathbf{40\ V}$

7. **C** Electrical theory

8. **C** $°F = (1.8 \times °C) + 32$
$°F = 1.8 \times 30 + 32$
$°F = 54 + 32 = 86$
$°F = \mathbf{86°F}$

9. **A** $°C = \frac{°F - 32}{1.8}$

$°C = \frac{104 - 32}{1.8}$

$°C = \frac{72}{1.8} = 40$

$°C = \mathbf{40°C}$

10. **D** Electrical theory

11. **B** *210.19(A), Informational Note No.3*
240 V × 0.03 = **7.2 V**

12. **B** $VD = \frac{Line\ Loss}{Supply\ Voltage} \times 100$

$VD = \frac{8}{115} \times 100 = 6.96$

$VD = \mathbf{6.96\%}$

13. **B** $E_{ff} = \frac{Output}{Input}$

$E_{ff} = \frac{295}{300} = 98$

$E_{ff} = \mathbf{98\%}$

14. **A** $I = \frac{P}{E}$

$I = \frac{6500}{115} = 56.62$

$P = E \times I$
$P = 120 \times 56.52 = 6782.4$
$P = 115 \times 56.52 = 6499.8$
$P = 6782.4 - 6499.8 = 282.6$
$P = \mathbf{283\ W}$
or
$P = E \times I$
$P = 5 \times 56.52 = 282.6$
$P = \mathbf{283\ W}$

15. **B** Electrical theory

16. **A** Electrical theory

17. **C** 60 m × 3.28084′ = **197′**

18. **A** 24″ × 2.54 cm = **60.96 cm**

19. **C** $R_2 = \frac{R}{N}$

$R_2 = \frac{12}{6} = 2$

$R_T = R_1 + R_2$

$R_T = 12 + 2 = \mathbf{14\ \Omega}$

20. **A** $R_p = \frac{1}{\frac{1}{R_1} + \frac{1}{R_2} + \frac{1}{R_3}}$

$R_p = \frac{1}{\frac{1}{5} + \frac{1}{7} + \frac{1}{9}}$

$R_p = \frac{1}{0.2 + 0.142 + 0.111}$

$R_p = \frac{1}{0.453} = 2.2$

$R_p = 2.2\ \Omega$
$R_T = 2.2\ \Omega + 8\ \Omega$
$R_T = \mathbf{10.2\ \Omega}$

Review Questions 2-3 _____33

1. **B** $R = \frac{P}{S}$

$R = \frac{120}{12} = 10$

$R = \mathbf{10{:}1}$

2. **D** $I = \frac{P}{E}$

$I = \frac{300}{12} = 25$

$I = \mathbf{25\ A}$

3. **C** $kVA = \frac{I_S \times E_S}{1000}$

$kVA = \frac{25 \times 12}{1000}$

$kVA = \frac{300}{1000} = 0.03$

$kVA = \mathbf{0.3\ kVA}$

4. **C** $I_P = \frac{kVA \times 1000}{E}$

$I_P = \frac{0.3 \times 1000}{120}$

$I_P = \frac{300}{120} = 2.5$

$I_P = \mathbf{2.5\ A}$

5. **B** $R = \frac{E_2}{P}$

$R = \frac{12 \times 12}{300}$

$R = \frac{144}{300} = 0.48$

$R = \mathbf{0.48\ \Omega}$

6. **C** Electrical theory

7. **A** Electrical theory
8. **C** Electrical theory
9. **C** Electrical theory
10. **C** The 120 V loads are in series across 240 V.
11. **A** $R = \frac{E_2}{P}$

 $R = \frac{120 \times 120}{25}$

 $R = \frac{14{,}400}{25} = 576$

 $R = \mathbf{576\ \Omega}$
12. **B** $R = \frac{E_2}{P}$

 $R = \frac{130 \times 130}{100}$

 $R = \frac{16{,}900}{100} = 169$

 $R = \mathbf{169\ \Omega}$
13. **A** $R_T = R_1 + R_2$

 $R_T = 576 + 169 = 745$

 $R_T = \mathbf{745\ \Omega}$
14. **C** $I = \frac{E}{R}$

 $I = \frac{240}{745} = 0.322$

 $I = \mathbf{0.322\ A}$
15. **D** $E = I \times R$

 $E = 0.322 \times 576 = 185.5$

 $E = \mathbf{185.5\ V}$
16. **A** $E = I \times R$

 $E = 0.322 \times 169 = 54.4$

 $E = \mathbf{54.4\ V}$

Review Questions 2-4 _____35

1. **C** *Primary*

 $P = I \times E$

 $P = 8.33 \times 120 = 999.6$

 $P = 999.6$ VA

 Secondary

 $P = I \times E$

 $P = 3.9 \times 240 = 936$

 $P = 936$ VA

 $E_{ff} = \frac{Output}{Input}$

 $E_{ff} = \frac{936}{99.6} = 93.6$

 $E_{ff} = \mathbf{94\%}$
2. **D** $E_{ff} = \frac{HP \times 746}{I \times E}$

 $E_{ff} = \frac{7.5 \times 746}{40 \times 240}$

 $E_{ff} = \frac{5595}{9600} = 0.58$

 $E_{ff} = \mathbf{58\%}$
3. **C** $E = I \times R$

 $E = 10 \times 58$

 $E = \mathbf{580\ V}$
4. **D** *Ch 9, Table 8:* 3 AWG = 52,620 CM

 210.19(A), Informational Note No.3

 240 V × 3% = 7.2 V

 Ch 9, Table 8: 1ϕ Cu = 25.8 Ω

 $I = \frac{VD \times CM}{D \times K}$

 $I = \frac{7.2 \times 52{,}620}{219 \times 25.8}$

 $I = \frac{378{,}864}{5650} = 67$

 $I = \mathbf{67\ A}$
5. **D** $R_T = \frac{1}{\frac{1}{R_1} + \frac{1}{R_2} + \frac{1}{R_3}}$

 $R_T = \frac{1}{\frac{1}{4} + \frac{1}{8} + \frac{1}{15}}$

 $R_T = \frac{1}{0.25 + 0.125 + 0.066}$

 $R_T = \frac{1}{0.441} = 2.27$

 $R_T = \mathbf{2.27\ \Omega}$
6. **A** $R_T = R_1 + R_2 + R_3$

 $R_T = 4 + 24 + 1 = 29$

 $R_T = \mathbf{29\ \Omega}$

 $I = \frac{E}{R}$

 $I = \frac{5}{1} = 5$

 $I = 5$ A

 $E = I \times R$

 $E = 5 \times 29 = 145$

 $E = \mathbf{145\ V}$
7. **D** $Input = \frac{Output}{E_{ff}}$

 $Input = \frac{5 \times 746}{0.80}$

 $Input = \frac{3730}{0.80} = 4662$

 $Input = \mathbf{4662\ VA}$
8. **B** $I = \frac{P}{E \times PF}$

 $I = \frac{10{,}000}{240 \times 0.85}$

 $I = \frac{10{,}000}{204} = 49$

 $I = \mathbf{49\ A}$
9. **D** $I = \frac{P}{E}$

 $I = \frac{15{,}000}{240} = 62.5$

 $I = \mathbf{62.5\ A}$
10. **D** $P = E \times I \times PF$

 $P = 230 \times 21 \times 0.93 = 4491.9$

 $P = \mathbf{4492\ VA}$
11. **B** $P = E \times I$

 $P = 220 \times 10 = 2200$

 $P = 2200$ W

 $Wh = W \times T$

 $Wh = 2200 \times 0.75 = 1650$

 $Wh = \mathbf{1650\ Wh}$
12. **B** $Cost = \frac{T \times W \times Cost / kWh}{1000}$

 $Cost = \frac{0.75 \times 2200 \times 0.12}{1000}$

 $Cost = \frac{198}{1000} = 0.198$

 $Cost = \mathbf{\$0.20}$
13. **A** $Input = \frac{Output}{E_{ff}}$

 $Input = \frac{500}{0.95} = 526$

 $Input = \mathbf{526\ VA}$
14. **A** $I = \frac{P}{E}$

 $I = \frac{2200}{12} = 18.3$

 $I = \mathbf{18.3\ A}$
15. **B** Electrical theory
16. **A** $I = \frac{E}{R}$

 $I = \frac{80}{10} = 8$

 $I = 8$ A

 $R = \frac{E}{I}$

 $R = \frac{40}{8} = 5$

 $R = \mathbf{5\ \Omega}$
17. **C** $R_T = R_1 + R_2 + R_3$

 $R_T = 6.5 + 0.193 + 0.193 = 6.88$

 $R_T = \mathbf{6.88\ \Omega}$
18. **C** $R_T = \frac{R_1}{N}$

 $R_T = \frac{4}{4} = 1$

 $R_T = \mathbf{1\ \Omega}$
19. **C** Trade knowledge
20. **C** Electrical theory

3 NEC® Examples

Review Questions 3-1 _____49

1. **B** *Table 310.15(B)(1)*
2. **B** Trade knowledge
3. **D** *430.6(A)(1)*
4. **A** *220.12*
5. **D** *Ch 9, Note 4*
6. **A** *Informative Annex C, Table C1*
7. **A** *Informative Annex C, Table C8*
8. **A** *Table 310.15(C)(1)*

9. **A** *Ch 9, Table 4*
10. **D** Trade knowledge
11. **D** *250.52(A)(3)*
12. **D** *Ch 9, Table 5*
13. **C** *Ch 9, Table 8*
14. **C** *430.22*
15. **A** *Table 310.16*
16. **A** *220.12*
17. **D** *700.10(D)*
18. **A** *324.41*
19. **B** *392.10(B)(1)(a)*
20. **D** *406.9(E)*

Review Questions 3-2 ______51

1. **C** *240.6(A)*
2. **D** *Ch 9, Table 8*
3. **C** *220.55; Table 220.55 Note 3*
   8750 W × 0.3 = 7000 W
   7000 W ÷ 240 V = 29.17A
4. **B** *Ch 9, Table 8*
5. **D** *Table 430.248:*
   5 HP, 208 V = 30.8 A
   FLC = **30.8 A**
6. **D** *430.52; Table 430.52:*
   30.8 A × 250% = 77 A
   240.6: Next higher standard size = 80 A
   ITCB = **80 A**
7. **A** *430.6(A)(2):* Nameplate rating = 28 A
   Overload Protection = **28 A**
8. **A** *Ch 9, Table 8:* 12 AWG stranded
   Cu = 1.98 Ω/1000′
   $\frac{1.98}{1000} = 0.00198$ Ω/ft
   $\frac{0.41}{0.00198} = 207'$
   Length = **207′**
9. **D** *Table 430.248:*
   3 HP = 17 A
   1 HP = 8 A
   430.62(A): 17 A × 250% = 42.5 A

	L1	N	L2
3 HP, 230 V	45	0	45
1 HP, 230 V	16	0	16
Total	61	0	61

   (not permitted to go to next higher size)
   240.6: 60 A
   ITCB = **60 A**
10. **B** *430.22:* 17 A × 125% = 21 A

	L1	N	L2
3 HP, 240 V	21	0	21
1 HP, 240 V	16	0	16
Total	37	0	37

   Ampacity = **37 A**
11. **B** *Table 310.16:*
   6 AWG RHW Al = 50 A
   Table 310.15(B)(2)(a):
   50 A × 75% = 37.5 A
   Ampacity = **37.5 A**
12. **D** *310.15(B)(3)(a) Ex 2, Table 310.15(B)(16):*
   OCPD = **30 A**
13. **C** *Table 220.55, Note 4; Table 220.55, Col. C:*
   One range = 8 kW
   Load = **8 kW**
14. **A** *430.6(A)(1); Table 430.248:*
   3 HP, 230 V = 17 A
   Ampacity = **17 A**
15. **A** *Table 220.55, Notes 1 & 2:*

8 × 14 kW	=	112 kW
5 × 8 kW	=	60 kW
10 × 12 kW	=	120 kW
Total	=	292 kW

   292 kW ÷ 23 = 12.7 kW
   (exceeds 12 kW by 1)
   23 ranges = 38 kW
   38 kW × 105% = 39.9 kW
   Demand Load = **39.9 kW**
16. **D** *Table 310.16 and Table 310.15(B)(1):*
   100 A × 58% = 58 A
   Ampacity = **58 A**
17. **C** *450.4(A)*
18. **B** *525.5(B)*
19. **A** *500.5(B)(1) Note No. 1, 2*
20. **D** *590.3(D)*

Review Questions 3-3 ______53

1. **B** *220.82(C)(2); Informative Annex D, Example D2(c):*
   P = *E* × *I*
   P = 30 × 240 = 7200
   P = 7200 VA
   7500 VA × 65% = 4875 VA
   7200 VA + 4875 VA = 12,075 VA
   Load = **12,075 VA**
2. **A** *220.82(C)(1) and (4)*
   P = *E* × *I*
   P = 45 × 240 = 10,800
   P = 10,800 VA
   15,000 VA × 65% = 9750 VA
   220.60: Smaller load = 9750 V (omit smaller load)
   Load = **10,800 VA**
3. **A** *220.12:*

General Lighting:		
1900 sq ft × 3	=	5700 VA
Small Appl and Laundry	=	4500 VA
Total	=	10,200 VA

   Table 220.42:

3000 × 100%	=	3000 VA
7200 VA × 35%	=	2520 VA
Total	=	5520 VA

   Neutral Load = **5520 VA**
4. **C** *Table 430.248:*
   5 HP = 28 A
   3 HP = 17 A
   Table 430.52:
   5 HP = 28 A × 250% = 70 A
   430.62(A): 70 A + 17 A = 87 A

	L1	N	L2
5 HP, 240 V	70	0	70
3 HP, 240 V	17	0	17
Total	87	0	87

   240.6: ITCB = **80 A**
5. **B** *430.24:*
   28 A × 125% = 35 A
   35 A + 17 A = 52 A

	L1	N	L2
5 HP, 230 V	35	0	35
3 HP, 230 V	17	0	17
Total	52	0	52

   Ampacity = **52 A**
6. **C** *Table 310.16:* 225 A requires 250 kcmil
   Conductors = **250 kcmil**
7. **B** *220.12:* 2200 sq ft × 3 VA = 6600 VA
   $I = \frac{VA}{V}$
   $I = \frac{6600}{120} = 55$
   55 A ÷ 15 A = 3.6 (round to 4)
   Circuits = **4**
8. **A** *Ch 9, Table 8:*
   Ω/kFT = per 1000 ft
   1.21 Ω ÷ 1000′ = 0.0021 Ω/ft
   0.00121 Ω × 90 = 0.1089 Ω
   Resistance = **0.1089** Ω
9. **B** *220.82(C)(1) and (4): Informative Annex D, Example D2(c):*
   A/C = 5760 W (omit smaller load)
   Heat = 10,000 VA × 65% = 6500 VA or 6.5 kVA
   Added Load = **6500 VA**
10. **C** *220.54:*
   Dryer load = 5000 VA
   $I = \frac{VA}{V}$
   $I = \frac{5000}{240} = 20.8$
   I = 20.8 A
   220.61(B): 20.8 A × 70% = 14.6 A
   Added Load = **14.6 A**
11. **C** *Informative Annex C, Table C2:*
   Eight 10 AWG in ½″
   Conductors = **8**
12. **A** *Table 220.55, Col B:*
   7000 VA × 80% = 5600 VA
   $I = \frac{VA}{V}$
   $I = \frac{5600}{240} = 23.3$
   I = **23 A**
13. **B** *Ch 9, Table 5A:*
   1 AWG = 0.415″
   Diameter = **0.415″**
14. **C** *Ch 9, Table 1:*
   3 conductors = 40% fill
   Ch 9, Table 4:
   3″ EMT conduit = 3.54
   Fill = **3.54 sq in.**

15. **C** *220.82(B)(3):*
Use nameplate rating
kW = **12 kW**

16. **B** *Table 220.55, Col A; 220.61(B):*
8000 W × 70% = 5600 W
or 5.6 kW
kW = **5.6 kW**

17. **C** *Table 550.31*
18. **C** *725.41(A)*
19. **D** *408.36(A)*
20. **A** *90.6*

Review Questions 3-4 _____55

1. **C** *Table 430.248; 430.6(A)(1):*
2 HP = 24 A
FLC = **24 A**

2. **B** *430.22:* 24 A × 125% = 30 A
Table 310.16: 10 AWG THWN = 30 A
Conductors = **10 AWG**

3. **D** *430.6; 430.32(A)(1):*
115% × 24 A = 27.6 A
Rating = **27.6 A**

4. **B** *430.32(A)(2):*
24 A × 140% = 33.6 A
Trip Current = **33.6 A**

5. **D** *Table 430.52:*
24 A × 250% = 60 A
ITCB = **60 A**

6. **A** *Table 430.52:*
24 A × 175% = 42 A
430.52(C)(1), Ex.1; 240.6:
Next higher standard size = 45 A
TDF = **45 A**

7. **D** *Table 430.52:*
24 A × 300% = 72 A
430.52(C)(1), Ex.1; 240.6:
Next higher standard size = 80 A
Fuse = **80 A**

8. **A** *Table 220.55, Col. C:*
Six ranges = 21 kW
Table 220.55, Note 1:
21 kW × 105% = 22 kW
Demand Load = **22 kW**

9. **D** *Table 310.15(C)(1):*
21–30 conductors = 45%
1 ÷ 45 = 2.2222
2.222 × 65 A = 144.4 A
Table 310.16: 1/0 AWG = 150 A
Conductors = **1/0 AWG**

10. **B** *Table 310.15(B)(1), Correction Factors:*
24°C = 105%
50 A × 105% = 52.5 A
Ampacity = **52.5 A**

11. **B** *Ch 9, Table 8:* 12 AWG = 1.98 Ω/kFT
1.98 ÷ 1000 = 0.00198 Ω/ft
0.00198 × 275′ = 0.5445 Ω
Resistance = **0.5445 Ω**

12. **C** *Ch 9, Table 8:*
250 kcmil = 0.0847 Ω/kFT
0.08476 ÷ 1000 = 0.0000847 Ω/ft
$R_T = \frac{R_1}{N} \times length$
$R_T = \frac{0.0000847}{2} \times length$
R_T = 0.0000423 × 175′ = 0.0074
R_T = 0.0074 Ω
Resistance = **0.0074 Ω**

13. **C** *Ch 9, Table 1:*
3 conductors = 40%
Fill = **40%**

14. **D** *Table 430.248:*
3 HP, 230 V, 1ϕ motor = 17 A
430.110(A):
17 A × 115% = 19.55 A
Rating = **20 A**

15. **A** *Table 220.55, Col. C:*
14 kW – 12 kW = 2 kW (or 10%)
Table 220.55, Note 1:
8000 × 5% = 400 VA
8000 + 400 + 400 = 8800 VA
or
8000 × 110% = 8800 VA
220.61(B): 8800 VA × 70% = 6160 VA
Neutral Load = **6160 VA**

16. **A** *Table 310.15(B)(1):*
6 AWG RH @ 113°F requires 0.82
65 A × 82% = 53.3 A
Ampacity = **53 A**

17. **D** *Table 310.15(B)(1):*
THHN @ 52°C requires 0.76
1 ÷ 0.76 = 1.3157
1.3157 × 50 A = 65.7 A
Table 310.16: Conductors = **6 AWG**

18. **B** *Table 220.12; 220.52(A)(B):*

General Lighting:	
2500 sq ft × 3 VA/sq ft	= 7500 VA
Small Appl and Laundry:	
1500 VA × 3	= 4500 VA
Total	12,000 VA

Load = **12,000 VA**

19. **A** *Table 220.55, Col B:*
6000 VA × 80% = 4800 VA
Load = **4800 VA**

20. **B** *Table 220.55, Col C:*
4 kW + 4 kW + 5 kW = 13 kW
Table 220.55, Col C;
Table 220.55, Note 1:
13 kW – 12 kW = 1 (or 5%)
Table 220.55, Note 4:
8000 VA × 5% = 400 VA
8000 VA + 400 VA = 8400 VA
Load = **8400 VA**

21. **B** *310.12(A), Table 310.12*
22. **D** *Table 250.66*
23. **B** *Table 310.15(B)(1)*
24. **C** *Informative Annex C, Table C11*
25. **D** *Table 310.15(C)(1)*
26. **A** *Ch 9, Table 5*
27. **B** *Table 250.66*
28. **A** *220.51*
29. **B** *Informative Annex C, Tables C1–C13A*
30. **D** 310.17(1); *Table 310.17*

4 Journeyman Electrician's Practice Licensing Exams (45 Minutes)
Practice Licensing Exams

Practice Licensing Exam 4-1 _____61

1. **B** *Table 314.16(A)*
2. **A** Trade knowledge
3. **A** Electrical theory
4. **A** Electrical theory
5. **C** *210.62*
6. **A** *310.10(G)(1)*
7. **B** *210.52(H)*
8. **C** *210.52(B)(1)(2), Ex. 1*
9. **B** *220.14(J)*
10. **A** *220.14(G)(2)*
11. **C** Electrical theory
12. **D** *300.4(G)*
13. **D** *250.52(A)(1) through (8)*
14. **A** *110.14(A)*
15. **B** *314.27(A)(2)*
16. **A** *430.6(A)(2)*
17. **A** *230.79(C)*
18. **B** Electrical theory
19. **A** *680.23(A)(4)*
20. **B** *518.1*

Practice Licensing Exam 4-2 _____63

1. **B** Electrical theory
2. **A** Electrical theory
3. **B** *225.26*
4. **C** *600.5(A)*
5. **B** Electrical theory
6. **A** *100*
7. **C** Electrical theory
8. **D** Electrical theory
9. **D** Trade knowledge
10. **A** Electrical theory
11. **D** Trade knowledge
12. **C** Electrical theory
13. **A** Trade knowledge
14. **A** Electrical theory
15. **A** Electrical theory
16. **C** Electrical theory
17. **C** Trade knowledge
18. **B** 890
19. **A** 890.2
20. **B** Electrical theory

Practice Licensing Exam 4-3 _____65

1. **C** *Table 310.104(A)*
2. **D** *300.22(A)*
3. **B** Electrical theory
4. **B** *250.136*
5. **D** *680.26(B)(7), Ex. 2*
6. **C** *480.2 Informational Note*

7. **C** *90.2(B)(2)*
8. **B** *Table 551.73(A)*
9. **D** *336.104*
10. **A** *210.19(A)(3)*
11. **D** *225.6(A)(1)*
12. **A** *225.7(C)*
13. **D** *230.71(B)*
14. **B** *Table 310.16*
15. **D** *250.52(A)(1) through (4)*
16. **C** *Table 310.4(A)*
17. **B** *334.80; Table 310.16 and 334.80*
18. **A** *Table 310.16*
19. **C** Electrical theory
20. **B** Electrical theory

Practice Licensing Exam 4-4 ______________ 67

1. **C** *300.5(G)*
2. **B** *370.20(A)(2)*
3. **B** *445.13(A)*
4. **B** Electrical theory
5. **D** Electrical theory
6. **C** Electrical theory
7. **B** *408.3(E)(1)*
8. **A** *314.21*
9. **B** *376.22(B)*
10. **B** *210.8(A)(9)*
11. **A** *240.51(A)*
12. **A** *Table 312.6(B)*
13. **C** *210.8(A)(10)*
14. **C** Trade knowledge
15. **C** *210.8(B)(1)*
16. **D** *210.52(C)(1)*
17. **C** *210.52(A)(3)*
18. **A** *220.43(B)*
19. **C** *Table 402.5*
20. **B** *342.14; 344.14*

Practice Licensing Exam 4-5 ______________ 69

1. **B** *230.24(B)(4)*
2. **B** *230.24(B)(2)*
3. **A** *450.21(A)*
4. **C** *334.23, 320.23(A)*
5. **B** Electrical theory
6. **A** Electrical theory
7. **A** *358.20(B)*
8. **D** *424.4(A)*
9. **D** *210.52(B)(1)*
10. **C** *Table 300.5*
11. **B** *230.82(4)*
12. **A** *110.14(B)*
13. **A** *250.24(C)*
14. **A** *230.53*
15. **C** *Table 300.19(A)*
16. **C** *210.12(A)*
17. **A** *300.6(D)*
18. **A** *210.52(C)(3)*
19. **A** *440.64*
20. **B** *250.70(1)*

Practice Licensing Exam 4-6 ______________ 71

1. **B** *700.12(C)*
2. **B** *225.6(A)(1)*
3. **C** *230.7, Ex. 1*
4. **A** *630.11(A)*
5. **B** *540.13*
6. **D** *110.34(C)*
7. **C** *240.30(A)(1)*
8. **C** *250.102(A)*
9. **D** *110.13(A)*
10. **A** *404.8(A)*
11. **D** *338.10(B)(1)*
12. **A** *Table 312.6(A)*
13. **C** *310.12, Table 310.12*
14. **C** *250.102(C)(1) Note 1*
15. **A** *100*
16. **A** *430.32(A)(1)*
17. **B** *382.30(A)*
18. **D** *230.51(A)*
19. **B** Trade knowledge
20. **C** Electrical theory

Practice Licensing Exam 4-7 ______________ 73

1. **D** Electrical theory
2. **B** Electrical theory
3. **C** Electrical theory
4. **A** Electrical theory
5. **B** Electrical theory
6. **A** *Table 300.5*
7. **C** *Table 250.122*
8. **D** Electrical theory
9. **C** Electrical theory
10. **A** *310.4(A)*
11. **A** *410.10(D)(1)*
12. **B** *314.28(A)(2)*
13. **C** *250.64(E)(1)*
14. **A** *250.64(A)(3)*
15. **B** *230.9(A), Ex.*
16. **C** *Table 250.66*
17. **D** *300.22(C)*
18. **A** *300.4(A)(1)*
19. **D** *110.15*
20. **B** Trade knowledge

Practice Licensing Exam 4-8 ______________ 75

1. **B** *100*
2. **C** Electrical theory
3. **A** Electrical theory
4. **D** Electrical theory
5. **A** Electrical theory
6. **C** Electrical theory
7. **A** Electrical theory
8. **C** Trade knowledge
9. **D** Electrical theory
10. **D** *410.56(E)*
11. **B** *424.35*
12. **C** *240.6(A)*
13. **D** Electrical theory
14. **C** 210.21(B)(1)
15. **C** *424.22(A), 210 Part II*
16. **B** *250.53(A)(2), Ex.*
17. **C** Electrical theory
18. **D** *Ch 9, Table 5*
19. **B** Electrical theory
20. **A** *Table 300.5*

Practice Licensing Exam 4-9 ______________ 77

1. **C** Electrical theory
2. **A** *422.16(B)(1)*
3. **D** *460.8(A)*
4. **A** Electrical theory
5. **A** *540.13*
6. **C** *310.3(C)*
7. **B** Electrical theory
8. **C** Electrical theory
9. **B** Electrical theory
10. **C** *210.12(A)(3)(b)*
11. **B** *90.6*
12. **A** *410.68*
13. **D** Electrical theory
14. **B** Electrical theory
15. **A** Electrical theory
16. **B** *550.2*
17. **D** *90.1(B)*
18. **C** Electrical theory
19. **A** *230.44*
20. **B** *250.102(C)(1)*

Practice Licensing Exam 4-10 ______________ 79

1. **B** *240.53(A)*
2. **B** Electrical theory
3. **D** *Table 220.12*
4. **A** Trade knowledge
5. **A** *344.42(B)*
6. **C** *500.5(C)*
7. **A** *250.140, Ex.*
8. **B** *Table 314.16(B)*
9. **D** Electrical theory
10. **A** *404.2(A)*
11. **A** Electrical theory
12. **A** *404.14(B)(2)*
13. **D** *Table 300.5*
14. **C** Electrical theory
15. **C** Trade knowledge
16. **A** *310.15(C)(1); Table 310.16*
17. **C** Electrical theory
18. **B** Electrical theory
19. **B** *310.10(G)(5)*
20. **C** *422.13*

Practice Licensing Exam 4-11 ______________ 81

1. **A** *358.42*
2. **B** *332.40(B)*
3. **D** *Art 100*
4. **A** *Table 310.15(C)(1)*
5. **C** *Table 314.16(A)*

6. **A** *250.22(3); 668.3(C)(3)*
7. **D** *404.10(B)*
8. **A** *Table 348.22*
9. **A** *210.4(C)*
10. **C** Electrical theory
11. **C** *300.7(A)*
12. **B** Electrical theory
13. **C** *408.18(A)*
14. **D** *110.13(B)*
15. **C** *314.16(C)(3)*
16. **B** *220.14(H)(2)*
17. **D** *230.6(1)*
18. **A** *314.29*
19. **D** *220.12*
20. **D** *230.82(1-11)*

Practice Licensing Exam 4-12 ______________ 83

1. **D** *250.53(A)(3)*
2. **C** *Table 300.19(A)*
3. **A** *352.12(D)*
4. **D** *Ch 9, Table 1*
5. **D** *240.10*
6. **B** *410.16(C)(2)*
7. **A** *422.16(B)(2)(1)*
8. **B** *220.82(C)(4)*
9. **C** *Table 310.16; 240.4(B)*
10. **D** *680.26(B)*
11. **C** Electrical theory
12. **D** Electrical theory
13. **C** *220.14(I)*
14. **D** *Table 220.56*
15. **D** *400.12(1) and (5)*
16. **D** Electrical theory
17. **D** Trade knowledge
18. **C** Trade knowledge
19. **C** Trade knowledge
20. **A** Electrical theory

Practice Licensing Exam 4-13 ______________ 85

1. **D** *Table 250.66:*
   Over 350 kcmil Cu–600 kcmil Cu = 1/0 AWG
   Cu GEC = **1/0 AWG**
2. **A** *630.11(A):* Nameplate primary current based on duty cycle.
   90 A × 55% = 49.5 or 50 A
   Supply Conductors = **50 A**
3. **B** *314.16; Tables 314.16(A) & (B):*

14 AWG	2 × 2 =	4.00 cu in.
12 AWG	3 × 2.25 =	6.75 cu in.
One EGC	=	2.25 cu in.
One receptacle	=	4.50 cu in.
Two cable clamps	=	2.25 cu in.
		19.75 cu in.

   Device Box = **19.75 cu in.**
4. **D** *314.16; Tables 314.16(A) & (B):*

12 AWG	4 × 2.25 =	9 cu in.
EGC	=	2.25 cu in.
		11.25 cu in.

   314.16(B)(1), Ex.: Less than five fixtures wires do not count.
   Device Box = **11.25 cu in.**
5. **C** *Table 310.16:*
   10 AWG THWN Cu = 35 A
   Table 310.15(B)(2)(a):
   32°C = 0.94
   35 A × 0.94 = 32.9 A
   Table 310.15(C)(1):
   32.9 A × 80% = 26.3 A
   Conductors = **26.3 A**
6. **A** $I = \frac{P}{E}$

   $I = \frac{1500}{120} = 12.5$

   I = 12.5 A
   Table 430.248:
   One 2 HP, 230 V motor = 12 A
   Table 430.248:
   Three 1 HP, 120 V motors = 16 A
   430.24: 12 A × 125% = 15 A

L1	N	L2
12.5	0	12.5
15	0	15
16	0	16
	16	16
43.5	16	59.5

   Neutral = **16 A**
7. **B** *348.22:* Two 12 AWG XHHW conductors permitted.
   *In addition, one uninsulated EGC of same sizeshall be permitted.
   Conduit = **fitting outside**
8. **C** Table *630.11(A):* 40 A × 78% = 31.2 A
   FLC = **31 A**
9. **D** *630.12(A):* 40 A × 200% = 80 A
   OCPD = **80 A**
10. **D** *460.8(A):* 18 A × 135% = 24.3 A
    Ampacity = **24.3 A**
11. **A** *Table 400.4*
12. **B** *440.4(C)*
13. **D** *600.5(C)(1)*
14. **D** *800.2*
15. **C** *404.2(B)*
16. **B** *Article 555.2*
17. **B** *410.153*
18. **A** *Table 326.24*
19. **A** *210.63(B)(1)*
20. **B** *518.1*

Practice Licensing Exam 4-14 ______________ 87

1. **D** *Table 310.16:*
   THWN Cu 200 A = 3/0 AWG
   Ungrounded Conductors = **3/0 AWG**
2. **B** R_3 is open.
   Circuit Fault = **opened**
3. **A** *314.16(B)(1), Ex.;*
   Table 314.16(A); Table 314.16(B):

12 AWG	2 × 2.25 =	4.5
12 AWG (EGC)	1 × 2.25 =	2.25
16 AWG (Omit)		
		6.75

   Table 314.16(A):
   Octagon Box = **4″ × 1¼″**
4. **B** R_1 is shorted.
   Circuit Fault = **shorted**
5. **D** *250.66(B):*
   GEC = **4 AWG**
6. **C** $PF = \frac{P}{I \times E}$

   $PF = \frac{5000}{30 \times 208}$

   $PF = \frac{5000}{6240} = 0.80$

   PF = **80%**
7. **A** *Table 314.16(A):*
   four AWG 12 phase conductors
   one AWG 12 neutral
   314.16(B)(5): 1 12 AWG EGC
   314.16(B)(1), Ex.: Fixture wires require no deduction
   Table 314.16(A):
   5 12 AWG requires 3″ × 2″ × 2¾″
   Device Box = **3″ × 2″ × 2¾″**
8. **D** *314.28(A)(1):* 8″ × 2″ = 16″
   Length = **16″**
9. **D** *314.16(A)(B); Table 314.16(A) & (B):*

14 AWG Ungrounded conductor	2 × 2 =	4
14 AWG Grounded conductor	2 × 2 =	4
14 AWG EGC		= 2
Total		= 10

   Box = **10 cu in.**
10. **C** *220.61(B)(2):*

250 A – 200 A	=	50 A
50 A × 70%	=	35 A
200 A + 35 A	=	235 A

    Load = **235 A**
11. **A** *800.100(A)(3)*
12. **A** *Table 300.19(A)*
13. **C** *225.26*
14. **D** *334.80*
15. **B** *Table 402.3*
16. **C** *620.11(A)*
17. **A** *424.22(B)*
18. **A** *445.14*
19. **D** *480.2, Informational Note*
20. **C** *490.2*

Practice Licensing Exam 4-15 ______________ 89

1. **D** *314.16; Table 314.16(A):*

Cable clamps	=	1 conductor
Six 12 AWG	=	6 conductors
SP switch	=	2 conductors
Total	=	9 conductors

   Table 314.16(A):
   Nine 12 AWG = 21
   Box = **21 cu in.**
2. **D** *Ch 9, Note 4; Ch 9, Table 4:*
   1.496 × 60% = 0.89
   Fill = **0.89 sq in.**
3. **B** *Ch 9, Table 5:*
   12 AWG TW6 × 0.0181 = 0.1086
   12 AWG THW6 × 0.0181 = 0.1086
   10 AWG THWN3 × 0.0211 = 0.0633
   Total = 0.2805
   Ch 9, Table 1:
   Over 2 conductors = 40%

Ch 9, Table 4: IMC = 0.384
Conduit = **1″**

4. **A** $I = W \div E$
$I = 10{,}000 \div 240 = 41.7$
I = **41.7 A**

5. **B** $I = W \div E$
$I = 10{,}000 \div 240 = 41.7$ A
$I = 41.7$ A
424.3(B): 41.7 A × 125% = 52 A
Table *240.6(A):* 60 A
OCPD = **60 A**

6. **C** *Table 310.16:*
THWN 300 kcmil = 285 A
285 A × 2 = 570 A
Table *240.6(A):* Standard ratings = 500 A, 600 A
240.4(B)(2):
THWN 300 kcmil
Conductors = **300 kcmil**

7. **A** *Ch 9, Table 8:*
300,000 CM × 2 = 600,000 CM (600 kcmil)
Table 250.66 : 1/0 AWG
GEC = **1/0 AWG**

8. **A** *310.12:*
Low-voltage wye is a 120/208 V system. 310.12 can be used for 120/240 V systems.
Table 310.12:
Feeder Conductors = **4 AWG**

9. **A** *Table 314.16(A):*
4¹¹⁄₁₆″ × 2⅛″ box
J-box = **4¹¹⁄₁₆″ × 2⅛″**

10. **A** *Ch 9, Table 8:*
1ϕ Cu = 25.8 Ω
210.19(A), Informational Note No. 3:
208 V × 3% = 6.24 V

$$CM = \frac{K \times I \times M \times D}{VD}$$

$$CM = \frac{12.9 \times 16 \times 2 \times 65}{6.24}$$

$$CM = \frac{68{,}112}{6.24} = 10{,}9153$$

Ch 9, Table 8:
12 AWG Cu = 6,530 CM
Conductors = **12 AWG Cu**

11. **A** *460.6(A)*
12. **D** *720.4*
13. **A** *660.5*
14. **B** *372.20*
15. **C** *455.2*
16. **A** *Table 514.3(B)(1)*
17. **B** *Article 820*
18. **A** *110.26(A)(4)(a)*
19. **B** *408.3(E)(1)*
20. **D** *393.10(3)*

Practice Licensing Exam 4-16 ____ 91

1. **B** *Table 250.122:*
EGC based on size of OCPD
150 A fuses require 6 AWG Cu
EGC = **6 AWG Cu**

2. **B** *250.122(B):*
EGCs shall be adjusted for VD:
8 AWG Cu
EGC = **8 AWG Cu**

3. **C** *240.4(D)(7):*
10 AWG limited to 30 A OCPD
OCPD = **30 A**

4. **C** *310.16:*
Conductor = **30 A**

5. **A** *Table 430.251(A):* ½ HP
Rating = **½ HP**

6. **B** *Ch 9, Table 9:* 0.062 Ω
Resistance = **0.062 Ω**

7. **A** $$Input = \frac{Output}{E_{ff}}$$

$$Input = \frac{500}{0.95} = 526$$

Power = **526 VA**

8. **C** *310.15(E)(12):*
Neutral = **current-carrying**

9. **D** *Table 430.250:*
25 HP motor = 53 A
Ampacity = **53 A**

10. **B** *220.14(G)(1):*
30′ × 200 W = 6000 W

$$I = \frac{P}{E}$$

$$I = \frac{6000}{120} = 50$$

$I = 50$ A

$$\frac{50\text{ A}}{15\text{ A}} = 3.3$$

Circuits = **4**

11. **A** *550.31(1)*
12. **C** *314.16(A)*
13. **C** *332.30*
14. **D** *530.2*
15. **A** *545.2*
16. **B** *552.10(E)(1); Table 552.10(E)(1)*
17. **B** *820.44(B)(1)*
18. **A** *Ch. 9, Table 1*
19. **A** *513.3(C)(1)*
20. **D** *490.2*

5 Journeyman Electrician's Practice Licensing Exams (180 minutes)
Practice Licensing Exams

Practice Licensing Exam 5–1 ____ 95

1. **A** *Table 220.12*
2. **B** *200.6(A)(2)*
3. **C** *100: location, dry;*
4. **D** *210.6(D)(1)(b)*
5. **C** *210.8(A)(6)*
6. **B** *210.50(C)*
7. **C** *210.52(H)*
8. **D** *210.8(A)(3), 210.63*
9. **A** *225.6(B)*
10. **A** Trade knowledge
11. **C** *410.10(F)*
12. **B** *517.2*
13. **A** *Table 310.4(A)*
14. **D** *314.27(C)*
15. **A** *Table 348.22*
16. **B** *408.3(E)(1)*
17. **D** *408.54*
18. **B** *314.21*
19. **D** *404.8(B)*
20. **A** *680.26(B)*
21. **D** *680.22(C)*
22. **D** *210.8(A)(1); 210.11(C)(3)*
23. **B** *110.14(C); Table 310.16*
24. **C** Electrical theory
25. **A** *210.63*
26. **D** *362.12(4)*
27. **D** *300.11(B)(1)*
28. **B** *352.28*
29. **D** *200.10(B)(1) and (2)*
30. **A** Electrical theory
31. **C** °*F* = (1.8 × °*C*) + 32
°*F* = 1.8 × 38 + 32 = 100.4°F
Reading = **100.4°F**

32. **B** *Table 430.248:*
2 HP = 12 A
3 HP = 17 A
5 HP = 28 A
Table 430.52:
28 A × 250% = 70 A
430.62:

2 HP	= 12 A
3 HP	= 17 A
5 HP	= 70 A
Total	= 99 A

240.6: Next lower standard size = 90 A
CB = **90 A**

33. **B** $$I = \frac{P}{E \times PF}$$

$$I = \frac{1200}{240 \times 0.65}$$

$$I = \frac{1200}{204} = 5.88$$

I = **5.88 A**

34. **D** *430.248:* 3 HP FLC = 34 A

$$E_{ff} = \frac{Output}{Input}$$

$$E_{ff} = \frac{3 \times 746}{34 \times 120}$$

$$E_{ff} = \frac{2238}{4080} = 55$$

E_{ff} = **55%**

35. **C** *250.66(B):* GEC = 4 AWG Cu
GEC = **4 AWG**

36. **B** *Table 430.248:*
2 HP, 208 V, 1ϕ = 13.2 A
Table 430.52:

13.2 A × 250% = 33 A
430.52(C)(1), Ex. 1:
35 A (Use next higher size.)
CB = **35 A**

37. **D** *Table 430.248:*
2 HP, 208 V, 1ϕ = 13.2 A
430.52(C)(1), Ex. 2(2):
13.2 A × 225% = 29.7 A
240.6: Standard size = 25 A
TDF = **25 A**

38. **D** *Table 430.248:*
2 HP, 208 V, 1ϕ = 13.2 A
430.52(C)(1), Ex. 2(3):
13.2 A × 400% = 52.8 A
240.6: Standard size = 50 A
CB = **50 A**

39. **C** *Table 220.12; 220.12:*
General Lighting:
2300 VA × 3 VA/sq ft = 6900 VA
Garage 150 sq ft (omit)

$$I = \frac{P}{E}$$

$$I = \frac{6900}{120} = 57.5$$

$I = 57.5$ A
57.5 A ÷ 15 A = 3.8
Circuits = **4**

40. **B** $P = I \times E$
$P = 35 \times 240 = 8400$
$P = 8400$ VA
Table 220.82(C)(3)
Amex D, Example D2(C):
8400 VA + 7500 VA = 13,275 VA
Added Load = **13,275 VA**

41. **D** 455.6(A)(1)
34 A × 125% = **42.5 A**

42. **B** *100*
43. **C** *Table 430.250*
44. **B** *110.26(A)(3)*
45. **D** *210.52(B)(2) Ex. 1*
46. **A** *230.9(A)*
47. **B** *240.5(B)(2)(1)*
48. **C** Electrical theory
49. **D** *240.6(A)*
50. **A** *Table 310.17*
51. **D** *Table 314.16(B)*
52. **A** *240.40*
53. **A** *344.26*
54. **B** *440.12(A)(1)*
55. **A** *680.21(A)(3)*
56. **C** Electrical theory
57. **C** *760.130(B)(2)*
58. **B** Electrical theory
59. **A** *250.68(C)(1)*
60. **D** *680.11(A)(1) to (7)*
61. **A** *430.233*
62. **B** *Table 430.248*
63. **B** NEMA Chart
64. **C** *250.52(A)(5)*
65. **D** *300.5(D)(1)*
66. **C** *314.17(8)(2)*
67. **C** *300.6(D)*
68. **A** *410.30(A)*
69. **C** *422.13*
70. **A** *410.116(A)(1)*
71. **C** *Table 402.5*
72. **B** *350.30(B)*
73. **C** *430.6(A)(1)*
74. **B** *501.15(A)(2)*
75. **A** *620.22(A)*
76. **D** *550.10(A)*
77. **B** *Table 514.3(B)(1)*
78. **A** *501.15(C)(3)*
79. **D** *680.23(F)(2)*
80. **C** *314.29*

Practice Licensing Exam 5–2 ______ 103

1. **C** *210.23(A)(1)*
2. **C** *250.118(7)(b)*
3. **B** *Table 344.30(B)(2)*
4. **A** *350.20(A)(B)*
5. **B** *300.14*
6. **C** *366.23(A)*
7. **B** *410.30(A)*
8. **C** *646.11(B)*
9. **A** *408.52*
10. **C** *225.6(A)(1)*
11. **B** *210.19(A)(1)(b)*
12. **C** *408.54*
13. **C** Electrical theory
14. **C** Electrical theory
15. **D** *324.41*
16. **B** *240.51(A)*
17. **A** *220.14(K)(2)*
18. **C** *200.10(B)(1)*
19. **B** *314.16(B)(2) and (3)*
20. **C** *230.53; 225.22*
21. **B** Electrical theory
22. **B** *Table 402.3*
23. **A** *705.11(C)(1)*
24. **C** *344.28*
25. **D** *550.16(A)(2)*
26. **C** *430.7(A)(1) through (9)*
27. **A** *300.5(D)(1)*
28. **B** *368.30*
29. **A** *386.12 (2)*
30. **B** *402.6*
31. **A** *Ch 9, Table 8:* 10 AWG Al = 10,380 CM
Ch 9, Table 8: 1ϕ Al = 42.4 Ω
210.19(A), Note 4:
240 V × 3% = 7.2 V

$$D = \frac{CM \times VD}{K \times I}$$

$$D = \frac{10{,}380 \times 7.2}{42.4 \times 15}$$

$$D = \frac{74{,}736}{636} = 117.5$$

D = **117.5′**

32. **B** *Table 430.248:* 32.2 A
FLC = **32.2 A**

33. **B** *Table 310.15(B)(1):*
53°C = 0.67
1 ÷ 0.67 = 1.492 (multiplier)
41 A × 1.492 = 61 A
Table 310.16: 61 A requires
6 AWG THWN Cu
Conductors = **6 AWG**

34. **D** *Table 310.16*
12 AWG THW Cu = 25 A
25 A × 0.80 = 20 A
Table 310.15(C)(1)
Ungrounded Conductors = **20 A**

35. **A** *Table 220.55, Note 1:*
15 kVA − 12 kVA = 3 kVA
3 kVA × 5% = 15%
Table 220.55, Col. C:
8000 W × 115% = 9200 W
Table 220.61(B):
9200 VA × 0.70 = 6440 VA
Load = **6440 VA**

36. **C** $$kVA = \frac{I_S \times E_S}{1000}$$

$$kVA = \frac{13 \times 240}{1000} = 3.12$$

Primary Rating = **3 kVA**

37. **B** *Ch 9, Table 8:*
8 AWG stranded Al = 1.28 Ω/kFT
1.28 Ω ÷ 1000 = 0.00128 Ω/ft
0.00128 Ω × 150′ = 0.192 Ω
Resistance = **0.192 Ω**

38. **C** *Ch 9, Note 4:*
1¼″ EMT 100% fill = 1.496 sq in.
Ch 9, Note 4:
1.496 sq in. × 0.60 = 0.8976 sq in.
Ch 9, Table 5:
6 AWG THHN = 0.0507 sq in.
0.8976 ÷ 0.0507= 17.7
Ch 9, Note 7: Drop 0.7
Conductors = **17**

39. **D** $R_T = R_1 + R_2 + R_3$
$R_T = 12 + 12 + 12 = 36$
R_T = **36 Ω**

40. **A** $$R_T = \frac{1}{\frac{1}{R_1} + \frac{1}{R_2} + \frac{1}{R_3}}$$

$$R_T = \frac{1}{\frac{1}{6} + \frac{1}{9} + \frac{1}{12}}$$

$$R_T = \frac{1}{0.166 + 0.111 + 0.083}$$

$$R_T = \frac{1}{0.36} = 2.78$$

R_T = **2.78 Ω**

41. **C** *Table 430.248:*
5 HP = 28 A
A/C: 28 × 240 = 6720
Heat: 5000 × 65% = 3250 VA
(omit smaller value than A/C)
Load = **6720 VA**

42. **A** *551.41(A)*
43. **B** *110.13(B)*
44. **A** *680.22(C)*

45. **B** *366.23:*
Thickness × Width = sq. inches
sq. inches × 1000 = Cu ampacity
0.5 × 4 = 2
2 × 1000 = 2000 A
Ampacity = **2000 A**

46. **C** *240.83(D)*
47. **A** *314.15*
48. **B** *366.30(B)*
49. **B** *430.111(B)(3)*
50. **A** *220.18(A)*
51. **C** *100*
52. **B** *220.55*
53. **C** $Input = \frac{output}{efficiency}$

$Input = \frac{500\ VA}{0.95} = 526\ VA$

Input = **526 VA**

54. **D** *230.22 Ex.*
55. **B** *300.11(B)(1) Ex.*
56. **B** *250.136*
57. **A** *430.42(C)*
58. **D** *338.100*
59. **C** *Table 220.54:*
6 × 6 kW × 75% = 27 kW
Load = **27 kW**
60. **A** *525.20(B)*
61. **B** *Table 310.4(A)*
62. **A** *680.22(B)(4)*
63. **C** *230.71(B)*
64. **D** *300.4(B)(1)*
65. **D** *225.19(B)*
66. **B** *250.28(A)*
67. **B** *100*
68. **A** *100*
69. **D** *100*
70. **C** *100*
71. **C** 400 ÷ 1000 = **0.4 kVA**
72. **B** Trade knowledge
73. **B** Electrical theory
74. **C** *Table 220.55:*
28 × 8 × 0.24 = 53.76 kW
Col. C:
15 + 28 = 43 kW
Load = **43 kW**

75. **A** *Ch 9, Table 8*
76. **C** *240.83(A)*
77. **C** *250.114(3)(b)*
78. **D** *Ch 9, Table 5*
79. **B** *210.23(A)(1)*
80. **A** *200.10(C)*

Practice Licensing Exam 5–3 ______________ 111

1. **A** *Table 402.5*
2. **A** *Table 400.4*
3. **C** *440.64*
4. **B** *Trade knowledge*
5. **D** *110.26(A)(3)*
6. **D** *300.4(A)(1)*
7. **A** *Informative Annex D, Examples*
8. **C** *450.21(A)*
9. **B** Trade knowledge
10. **C** *352.12(B)*
11. **B** *344.20(B)*
12. **D** Electrical theory
13. **A** Electrical theory
14. **A** *240.4(D)(4)*
15. **B** Electrical theory
16. **C** *230.72(A)*
17. **D** *590.4(F)*
18. **B** *250.122(B); Table 250.122*
19. **C** *250.52(B)(1)*
20. **D** *340.12(1)*
21. **B** *410.115(A)*
22. **B** *250.102(E)(2)*
23. **D** *376.56(A)*
24. **A** *230.54(A)*
25. **B** *445.13(A)*
26. **A** *Table 310.3*
27. **A** *Ch 9, Table 4 (Article 352)*
28. **A** *320.12(2)*
29. **B** *408.54*
30. **C** *728.5(F)*
31. **A** *Table 430.248:*
2 HP, 240 V motor = 12 A
FLC = **12 A**

32. **D** *430.22:* 12 A × 125% = 15 A
Table 310.16:
THWN Cu = 14 AWG
Wire = **14 AWG**

33. **B** *430.32(A)(1):*
12 A × 115% = 13.8 A
Ampacity = **13.8 A**

34. **B** Table *430.52:* 12 A × 300% = 36 A
430.52(C)(1):
Use next higher standard size
240.6: 40 A
NTDF = **40 A**

35. **A** *Ch 9, Table 8:* 14 AWG Cu
= 4110 CM
Ch 9, Table 8: 1ϕ Cu = 25.8 Ω
210.19(A), Note 4:
240 V × 3% = 7.2 V

$D = \frac{CM \times VD}{K \times I}$

$D = \frac{4110 \times 7.2}{25.8 \times 12}$

$D = \frac{29{,}592}{309.6}$

D = **95.6**

36. **D** *Ch 9, Table 8:*
14 AWG solid uncoated =
3.07 Ω/kFT
$\frac{3.07}{1000} = 0.00307\ \Omega/ft$
0.00307 Ω × 95.6 = 0.2935 Ω
R_T = **0.2935 Ω**

37. **B** *Output* = HP × 746
Output = 2 HP × 746 W = 1492 VA
Output = **1492 VA**

38. **C** *Table 220.55, Col. C:*
15 kW + 28 kW = 43 kW
Demand Load = **43 kW**

39. **A** *Table 220.55, Note 4:*
Nameplate rating = 3000 W
Branch Circuit Load = **3000 VA**

40. **C** $R_T = \frac{R_1}{N}$

$R_T = \frac{8}{2}$

$R_T = 4$
$R_T = R_1 + R_2$
$R_T = 4 + 4$
R_T = **8 Ω**

41. **A** Trade knowledge
42. **D** *110.5*
43. **D** Trade knowledge
44. **D** *250.119*
45. **A** Trade knowledge
46. **A** Trade knowledge
47. **B** Trade knowledge
48. **A** Trade knowledge
49. **B** Trade knowledge
50. **D** Trade knowledge
51. **C** Trade knowledge
52. **A** *100*
53. **D** *100*
54. **D** *100*
55. **A** *100*
56. **B** *Table 220.55, Note 1:*
15 kW – 12 kW = 3 kW
3 × 0.05 = 15%
8000 × 1.15 = 9200 W
9200 W ÷ 240 V = 38 A
210.19(A)(3):
Rating = **40 A**

57. **D** Trade knowledge
58. **C** *100*
59. **B** *406.4(A)*
60. **B** Trade knowledge
61. **C** *220.60:*
5 HP × 240 V = 6720 VA
Heat = 5000 VA
220.21:
Load = **6720 VA**

62. **C** *300.13(B)*
63. **D** *400.14, Informational Note*
64. **D** Trade knowledge
65. **A** *Table 220.55:*
28 × 6 kW × 24% = 40.3 kW
Load = **40.3 kW**

66. **A** *Table 430.248:*

$CM = \frac{K \times I \times D}{V \times D}$

$CM = \frac{25.8 \times 40 \times 350}{240 \times 0.03}$

$CM = \frac{361{,}200}{7.2}$

CM = 50,167 CM
Ch. 9, *Table 8:*
52,620 CM = 3 AWG Cu
Table 310.16:
The lowest rated wire is 85 A
430.22:
40 A × 125% = 50 A
3 AWG

67. **A** *366.23(A):*
4 × 0.5 = 2 sq in.
2 × 1000 = 2000 A
Ampacity = **2000 A**

68. **D** *Table 314.16(A)*
69. **C** Trade knowledge
70. **A** *550.13(D)*
71. **B** Trade knowledge
72. **C** *500.7(I)*
73. **C** *314.16(C)(2)*
74. **C** *250.118(7)*
75. **A** *240.4(D)(5)*
76. **A** *Table 430.248; 430.32(A)(1)*
77. **A** *240.54(C)*
78. **C** *Table 220.55, Note 2*
79. **B** *Tables 310; 310.4(A)*
80. **D** *406.7(A)*

Practice Licensing Exam 5–4 ______________ 119

1. **A** Electrical theory
2. **C** *220.14(J)(1)*
3. **D** *311.12(B)*
4. **D** Trade knowledge
5. **A** *240.4(B); 310.16*
6. **A** *240.5(B)(4)*
7. **D** *392.10(A); Table 392.10(A)*
8. **A** *372.20*
9. **D** *630.12(A)*
10. **B** *Table 250.66*
11. **C** *410.5, Ex.*
12. **C** *240.6(A)*
13. **A** *250.53(A)(3)*
14. **A** *310.3(C)*
15. **D** *230.28*
16. **A** *590.3(B)*
17. **A** *210.52(C)*
18. **B** *210.11(C)(1)*
19. **D** *334.10*
20. **A** *90.2(A)*
21. **D** *250.66*
22. **C** Electrical theory
23. **A** *358.26*
24. **C** *430.22*
25. **D** *430.109(B)*
26. **A** *408.18(A)*
27. **A** *210.52(A)(2)(1)*
28. **C** *Ch 9, Table 4*
29. **B** *Table 314.16(A)*
30. **B** *555.33(A)(4)*
31. **D** *Table 220.55, Notes 1 & 2:*
10 × 12 kW = 120 kW
8 × 14 kW = 112 kW
18 ranges = 232 kW
232 kW ÷ 18 = 12.88 kW (average)
Table 220.55, Col. (C):
18 ranges = 33 kW
33 kVA × 1.05 = 34.65 kVA
Demand Load = **34.65 kVA**

32. **A** *Table 220.55:*
16 × 6 kVA × 0.28 = 26.9 kVA
Demand Load = **26.9 kVA**

33. **B** *Table 430.247:*
5 HP, 240 V, DC motor FLC = 20 A
FLC = **20 A**

34. **B** $P = I \times E$
$P = 4 \times 230 = 920$
$P = 920$ VA
Losses = Input – Output
Losses = 920 – 746 = 174
Losses = **174 VA**

35. **C** *Table 220.55, Col. C:*
Nine 10 kW ranges = 24 kW
Demand Load = **24 kVA**

36. **A** $R_T = \dfrac{1}{\dfrac{1}{R_1}+\dfrac{1}{R_2}+\dfrac{1}{R_3}}$

$R_T = \dfrac{1}{\dfrac{1}{3}+\dfrac{1}{6}+\dfrac{1}{9}}$

$R_T = \dfrac{1}{0.333+0.166+0.111}$

$R_T = \dfrac{1}{0.61} = 1.639$

R_T = **1.64 Ω**

37. **B** $R_T = R_1 + R_2 + R_3$
$R_T = 3 + 6 + 9 = 18\ \Omega$
R_T = **18 Ω**

38. **A** *Table 310.16:*
10 AWG THWN Cu = 35 A
Table 310.15(C)(1);
Table 310.15(B)(1):
35 × 0.80 × 82% = 22.96 A
Ampacity = **23 A**

39. **C** $P = I \times E$
$P = 28 \times 240 = 6720$
$P = 6720$ VA
Informative Annex D, Example D2(C):
7500 VA × 65% = 4875 VA
220.82(C)(3):
6720 VA + 4875 VA = 11,595 VA
Added VA = **11,595 VA**

40. **B** *Ch 9, Table 8:* $K = 25.8\ \Omega$
210.19(A), Note 4:
120 V × 3% = 3.6 V

$D = \dfrac{K \times I \times D}{VD}$

$CM = \dfrac{25.8 \times 12 \times 116}{3.6}$

$CM = \dfrac{35{,}913.6}{3.6} = 9976$ CM

Ch 9, Table 8: 10 AWG Cu
Conductors = **10 AWG Cu**

41. **D** Trade knowledge
42. **B** Trade knowledge
43. **A** Trade knowledge
44. **B** *110.13(A)*
45. **B** *Table 220.55, Note 2:*
5 kW + 4 kW = 9 kW
9 kW × 65% = 5850 kW
Demand Load = **5850 kW**

46. **A** $kVA = \dfrac{I_S \times E_S}{1000} = 0.3$ kVA

$I_S = \dfrac{0.3 \times 1000}{12\ \text{V}} = 2.5$ A

Primary current = **2.5 A**

47. **C** *338.10(B)(4)(1)*

48. **A** $R = \dfrac{P}{S}$

$R = \dfrac{120\ \text{V}}{12\ \text{V}} = 10$

10:1

49. **B** *250.138(A)*
50. **C** *300.5(G)*
51. **A** Trade knowledge
52. **B** *334.23 and 320.23(A)*
53. **A** *225.19(B)*
54. **B** *240.4(D)(5), 240.92(C)(2)*
55. **C** *100*
56. **D** *240.83(A–E)*
57. **A** *680.22(B)(3)*
58. **D** *352.12*
59. **D** *250.70*
60. **D** *680.26(B)*
61. **A** *250.53(B)*
62. **B** Electrical theory
63. **B** *Table 310.15(C)(1)*
64. **A** *220.61(A):*
The neutral shall carry the maximum unbalanced load.

65. **A** *210.11(B)*

66. **C** $kVA = \dfrac{I_S \times E_S}{1000}$

$kVA = \dfrac{500}{1000}$

kVA = **0.5 kVA**

67. **D** *220.18(A); 440.6(A)*
68. **D** *300.10 Ex. 1*
69. **D** *100; Part III*
70. **D** *310.10(6)(1)*
71. **B** Trade knowledge
72. **C** Trade knowledge
73. **D** *210.21(A)*
74. **A** *Table 220.55:*
8000 + (400 + 400) = 8800 kW
or 8000 × 1.10 = 8.8 kW
Demand Load = **8.8 kW**

75. **A** *220.61:*
6400 × 70% = 4.48 kW
Neutral Load = **4.48 kW**

76. **D** *220.61(A):*
6400 – 5400 = 1000 VA
1000 VA ÷ 120 V = **8.33 A**

77. **A** *220.54:*
5000 × 2 = **10 kVA**

78. **C** *100*
79. **C** Trade knowledge
80. **C** *430.24; Table 430.248*
20 × 1.25 + 20 = **45 A**

Appendix

DECIMAL EQUIVALENTS OF AN INCH							
Fraction	Decimal	Fraction	Decimal	Fraction	Decimal	Fraction	Decimal
1/64	0.015625	17/64	0.265625	33/64	0.515625	49/64	0.765625
1/32	0.03125	9/32	0.28125	17/32	0.53125	25/32	0.78125
3/64	0.046875	19/64	0.296875	35/64	0.546875	51/64	0.796875
1/16	0.0625	5/16	0.3125	9/16	0.5625	13/16	0.8125
5/64	0.078125	21/64	0.328125	37/64	0.578125	53/64	0.828125
3/32	0.09375	11/32	0.34375	19/32	0.59375	27/32	0.84375
7/64	0.109375	23/64	0.359375	39/64	0.609375	55/64	0.859375
1/8	0.125	3/8	0.375	5/8	0.625	7/8	0.875
9/64	0.140625	25/64	0.390625	41/64	0.640625	57/64	0.890625
5/32	0.15625	13/32	0.40625	21/32	0.65625	29/32	0.90625
11/64	0.171875	27/64	0.421875	43/64	0.671875	59/64	0.921875
3/16	0.1875	7/16	0.4375	11/16	0.6875	15/16	0.9375
13/64	0.203125	29/64	0.453125	45/64	0.703125	61/64	0.953125
7/32	0.21875	15/32	0.46875	23/32	0.71875	31/32	0.96875
15/64	0.234375	31/64	0.484375	47/64	0.734375	63/64	0.984375
1/4	0.250	1/2	0.500	3/4	0.750	1	1.000

FORMULAS . . .

EFFICIENCY

$$E_{ff} = \frac{\text{Output}}{\text{Input}}$$

$$\text{Input} = \frac{\text{Output}}{E_{ff}}$$

$$\text{Output} = \frac{\text{Input}}{E_{ff}}$$

$$E_{ff} = \frac{HP \times 746}{I \times E}$$

$$HP = \frac{I \times E \times E_{ff}}{746}$$

$$I = \frac{HP \times 746}{E \times E_{ff}}$$

POWER FACTOR
(Watts/Meter Reading)

$$PF = \frac{P}{E \times I}$$

$$PF = \frac{P_T}{P_A}$$

$$PF = \frac{R}{Z}$$

POWER
(SINGLE-PHASE)

$$P = E \times I$$

$$P = I^2 \times R$$

$$P = \frac{E^2}{R}$$

$$E = \sqrt{P \times R}$$

$$E = \frac{P}{I \times PF}$$

$$I = \frac{P}{E \times PF}$$

$$I = \sqrt{\frac{P}{R}}$$

$$R = \frac{E^2}{P}$$

$$R = \frac{P}{I^2}$$

OHM'S LAW

$$E = I \times R$$

$$R = \frac{E}{I}$$

$$I = \frac{E}{R}$$

HORSEPOWER

1 HP = 746 watts

BUSBAR CAPACITY

thickness × width = square inches

square inches × 1000 = Cu ampere ampacity

square inches × 700 = AL ampere ampacity

ENERGY COST

$$\text{cost} = \frac{\text{watts} \times \text{hours used} \times \text{cost per kWh}}{1000}$$

SERIES CIRCUITS

$$E_T = E_1 + E_2 + E_3 + \ldots$$

$$I_T = I_1 = I_2 = I_3 = \ldots$$

$$R_T = R_1 + R_2 + R_3 + \ldots$$

PARALLEL CIRCUIT RESISTANCE

$$R_T = \frac{1}{\frac{1}{R_1} + \frac{1}{R_2} + \frac{1}{R_3}}$$

$$R_T = \frac{R_1 \times R_2}{R_1 + R_2}$$

$$R_T = \frac{R_1}{N}$$

PARALLEL CIRCUIT
VOLTAGE AND CURRENT

$$E_T = E_1 = E_2 = E_3 = \ldots$$

$$I_T = I_1 + I_2 + I_3 + \ldots$$

TEMPERATURE
CONVERSION

$$°F = (1.8 \times °C) + 32$$

$$°C = \frac{°F - 32}{1.8}$$

VOLTAGE DROP

$$VD = \frac{K \times I \times D}{CM}$$

$$CM = \frac{K \times I \times D}{VD}$$

$$I = \frac{VD \times CM}{D \times K}$$

$$D = \frac{CM \times VD}{K \times I}$$

$$VD = \frac{\text{line loss}}{\text{supply voltage}} \times 100$$

TRANSFORMERS
SINGLE-PHASE

$$kVA = \frac{I_P \times E_P}{1000}$$

$$kVA = \frac{I_S \times E_S}{1000}$$

$$I_P = \frac{kVA \times 1000}{E_P}$$

$$I_S = \frac{I_P \times E_P}{E_S}$$

TRANSFORMER POWER
(100% EFFICIENT)

$$E_P \times I_P = E_S \times I_S$$

TRANSFORMER TURNS

$$T_P = \frac{I_S \times T_S}{I_P}$$

$$T_S = \frac{E_s \times T_P}{E_P}$$

. . . FORMULAS

TRANSFORMER TURNS (E AND I)

$$E_S = \frac{E_P \times T_S}{T_P}$$

$$E_P = \frac{E_S \times T_P}{T_S}$$

$$T_P = \frac{E_P \times T_S}{E_S}$$

$$I_P = \frac{I_S \times T_S}{T_P}$$

$$\frac{I_P}{I_S} = \frac{T_S}{T_P}$$

TRANSFORMER RATIO

$$ratio = \frac{E_P}{E_S}$$

Note: Primary number is shown first.

A/C MOTORS

$$RPM = \frac{F \times 120}{P}$$

$$F = \frac{RPM \times P}{120}$$

$$P = \frac{F \times 120}{RPM}$$

CONVERSIONS

1 inch = 25.4 millimeters
1 inch = 2.54 centimeters
1 foot = 0.3048 meters
1 millimeter = 0.03937 inch
1 centimeter = 0.3937 inch
1 meter = 3.28084 feet
1 cubic inch = 16.387064 cubic centimeters
1 cubic centimeter = 0.061024 cubic inch
1 square millimeter = 0.001550 square inch
1 square inch = 645.16 square millimeters

POWER/THREE-PHASE

$$P = I \times E \times 1.73 \times PF$$

Note: Do not use 1.73 for single-phase.

$$E = \frac{P}{I \times 1.73 \times PF}$$

$$I = \frac{P}{E \times 1.73 \times PF}$$

$$PF = \frac{P}{E \times I \times 1.73}$$

$$kVA = \frac{I_P \times E_P \times 1.73}{1000}$$

$$kVA = \frac{I_S \times E_S \times 1.73}{1000}$$

$$I = \frac{kVA \times 1000}{1.73 \times E \text{ (primary or secondary delta or wye)}}$$

BALANCED THREE-PHASE WYE

Line E = 208 V

Phase E = 120 V

$$I_{line} = I_{phase}$$

$$E_{line} = E_{phase} \times 1.73$$

$$E_{phase} = \frac{E_{line}}{1.73}$$

$$I_{phase} = \frac{\text{VA per phase}}{\text{volts per phase}}$$

BALANCED THREE-PHASE DELTA

Line E = 240 V

Phase E = 240 V

$$I_{phase} = \frac{\text{phase VA}}{\text{phase volts}}$$

$$I_{line} = 1.73 \times I_{phase}$$

BALANCED THREE-PHASE DELTA OR WYE

total watts = $1.73 \times E$ line $\times$ I Line

$$\frac{\text{total watts}}{3} = \text{watts per phase}$$

RESIDENTIAL ELECTRICAL SYMBOLS . . .

LIGHTING OUTLETS

Description	Symbol
OUTLET BOX AND INCANDESCENT LIGHTING FIXTURE	CEILING WALL
INCANDESCENT TRACK LIGHTING	
BLANKED OUTLET	B B
DROP CORD	D
EXIT LIGHT AND OUTLET BOX. SHADED AREAS DENOTE FACES.	
OUTDOOR POLE-MOUNTED FIXTURES	
JUNCTION BOX	J J
LAMPHOLDER WITH PULL SWITCH	L PS L PS
MULTIPLE FLOODLIGHT ASSEMBLY	
EMERGENCY BATTERY PACK WITH CHARGER	B
INDIVIDUAL FLUORESCENT FIXTURE	
OUTLET BOX AND FLUORESCENT LIGHTING TRACK FIXTURE	
CONTINUOUS FLUORESCENT FIXTURE	
SURFACE-MOUNTED FLUORESCENT FIXTURE	

PANELBOARDS

Description	Symbol
FLUSH-MOUNTED PANELBOARD AND CABINET	
SURFACE-MOUNTED PANELBOARD AND CABINET	

CONVENIENCE OUTLETS

Description	Symbol
SINGLE RECEPTACLE OUTLET	
DUPLEX RECEPTACLE OUTLET	
TRIPLEX RECEPTACLE OUTLET	
SPLIT-WIRED DUPLEX RECEPTACLE OUTLET	
SPLIT-WIRED TRIPLEX RECEPTACLE OUTLET	
SINGLE SPECIAL-PURPOSE RECEPTACLE OUTLET	
DUPLEX SPECIAL-PURPOSE RECEPTACLE OUTLET	
RANGE OUTLET	R
SPECIAL-PURPOSE CONNECTION	DW
CLOSED-CIRCUIT TELEVISION CAMERA	
CLOCK HANGER RECEPTACLE	C
FAN HANGER RECEPTACLE	F
FLOOR SINGLE RECEPTACLE OUTLET	
FLOOR DUPLEX RECEPTACLE OUTLET	
FLOOR SPECIAL-PURPOSE OUTLET	
UNDERFLOOR DUCT AND JUNCTION BOX FOR TRIPLE, DOUBLE, OR SINGLE DUCT SYSTEM AS INDICATED BY NUMBER OF PARALLEL LINES	

BUSDUCTS AND WIREWAYS

Description	Symbol
SERVICE, FEEDER, OR PLUG-IN BUSWAY	B B B
CABLE THROUGH LADDER OR CHANNEL	C C C
WIREWAY	W W W

SWITCH OUTLETS

Description	Symbol
SINGLE-POLE SWITCH	S
DOUBLE-POLE SWITCH	S_2
THREE-WAY SWITCH	S_3
FOUR-WAY SWITCH	S_4
AUTOMATIC DOOR SWITCH	S_D
KEY-OPERATED SWITCH	S_K
CIRCUIT BREAKER	S_{CB}
WEATHERPROOF CIRCUIT BREAKER	S_{WCB}
DIMMER	S_{DM}
REMOTE CONTROL SWITCH	S_{RC}
WEATHERPROOF SWITCH	S_{WP}
FUSED SWITCH	S_F
WEATHERPROOF FUSED SWITCH	S_{WF}
TIME SWITCH	S_T
CEILING PULL SWITCH	(S)
SWITCH AND SINGLE RECEPTACLE	S
SWITCH AND DOUBLE RECEPTACLE	S
A STANDARD SYMBOL WITH AN ADDED LOWERCASE SUBSCRIPT LETTER IS USED TO DESIGNATE A VARIATION IN STANDARD EQUIPMENT.	a.b a.b $S_{a.b}$

. . . RESIDENTIAL ELECTRICAL SYMBOLS

COMMERCIAL AND INDUSTRIAL SYSTEMS

Description	Symbol
PAGING SYSTEM DEVICE	
FIRE ALARM SYSTEM DEVICE	
COMPUTER DATA SYSTEM DEVICE	
PRIVATE TELEPHONE SYSTEM DEVICE	
SOUND SYSTEM	
FIRE ALARM CONTROL PANEL	FACP

SIGNALING SYSTEM OUTLETS FOR RESIDENTIAL SYSTEMS

Description	Symbol
PUSHBUTTON	
BUZZER	
BELL	
BELL AND BUZZER COMBINATION	
COMPUTER DATA OUTLET	
BELL RINGING TRANSFORMER	BT
ELECTRIC DOOR OPENER	D
CHIME	CH
TELEVISION OUTLET	TV
THERMOSTAT	T

UNDERGROUND ELECTRICAL DISTRIBUTION OR ELECTRICAL LIGHTING SYSTEMS

Description	Symbol
MANHOLE	M
HANDHOLE	H
TRANSFORMER-MANHOLE OR VAULT	TM
TRANSFORMER PAD	TP
UNDERGROUND DIRECT BURIAL CABLE	
UNDERGROUND DUCT LINE	
STREET LIGHT STANDARD FED FROM UNDERGROUND CIRCUIT	

ABOVEGROUND ELECTRICAL DISTRIBUTION OR LIGHTING SYSTEMS

Description	Symbol
POLE	
STREET LIGHT AND BRACKET	
PRIMARY CIRCUIT	
SECONDARY CIRCUIT	
DOWN GUY	
HEAD GUY	
SIDEWALK GUY	
SERVICE WEATHERHEAD	

PANEL CIRCUITS AND MISCELLANEOUS

Description	Symbol
LIGHTING PANEL	
POWER PANEL	
WIRING – CONCEALED IN CEILING OR WALL	
WIRING – CONCEALED IN FLOOR	
WIRING EXPOSED	
HOME RUN TO PANEL BOARD Indicate number of circuits by number of arrows. Any circuit without such designation indicates a two-wire circuit. For a greater number of wires, indicate as follows: (3 wires) (4 wires), etc.	
FEEDERS Use heavy lines and designate by number corresponding to listing in feeder schedule.	
WIRING TURNED UP	
WIRING TURNED DOWN	
GENERATOR	G
MOTOR	M
INSTRUMENT (SPECIFY)	I
TRANSFORMER	T
CONTROLLER	
EXTERNALLY-OPERATED DISCONNECT SWITCH	
PULL BOX	

ABBREVIATIONS . . .

A

abbreviation ABBR
acoustic ACST
acoustical tile ACT. or AT.
adhesive ADD. or ADH
adjustable ADJ
adjustable-trip circuit breaker ATCB
aggregate AGGR
aileron AIL
air conditioner A/C
air handler A/H
airtight AT
alarm ALM
alloy ALY
alternating current AC
aluminum Al
ambient AMB
Ambulatory Health Care Center AHCC
American National Standards Institute ANSI
American Wire Gauge AWG
ammeter A or AM
ampere A or AMP
ampere interrupting rating AIR.
amps A
anchor bolt AB
anode A
antenna ANT.
apartment APT
appliance APPL
approved APPD or APVD
approximate APPROX
approximately APPROX
architectural ARCH.
architecture ARCH.
area A
area drain AD
armature A or ARM.
asphalt ASPH
asphalt tile AT.
as required AR
Assured Equipment Grounding Program AEGP
astragal A
Authority Having Jurisdiction AHJ
automatic AU or AUTO
automatic sprinkler AS
auxiliary AUX
avenue AVE
azimuth AZ

B

basement BSMT
bathroom B
bathtub BT
battery (electric) BAT.
beam BM
bearing BRG
bearing plate BPL or BRG PL
bedroom BR
benchmark BM
black BK
block BLK
blocking BLKG
blue BL
board BD
board foot BF
bonding jumper BJ
boulevard BLVD
brake relay BR
brass BRS
brick BRK
bridge BRDG
bronze BRZ
brown BR
building BL or BLDG
building line BL
built-up roofing BUR
bypass BYP

C

cabinet CAB.
cable CA
Cable Antenna Television CATV
cantilever CANTIL
capacitor CAP.
cased opening CO
casement CSMT
casing CSG
cast iron CI
cast-iron pipe CIP
cast steel CS
cast stone CS or CST
catch basin CB
cathode K
ceiling CLG
cellar CEL
Celsius °C
cement CEM
cement floor CF
center CTR
centerline CL
center-to-center C to C
centigrade C
central processing unit CPU
ceramic CER
ceramic tile CT
ceramic-tile floor CTF
channel CHAN
chapter CH
chimney CHM
circuit CIR or CKT
circuit breaker CB
circuit interrupter CI
circular mils CM
cleanout CO
clockwise CW
closet CLO
coarse CRS
coated CTD
coaxial COAX
Code-Making Panel CMP
cold air CA
cold water CW
column COL
compacted COMP
concrete CONC
concrete block CCB or CONC BLK
concrete floor CCF
concrete pipe CP
condenser COND
conductor resistivity K
conduit C or CND
construction joint CJ
continuous CONT
contour CTR
control CONT
control joint CJ or CLJ
control relay CR
control relay master CRM
copper Cu
corner COR
cornice COR
corrugated CORR
counterclockwise CCW
counter electromotive force CEMF
county CO
cubic CU
cubic foot CU FT
cubic foot per minute CFM
cubic foot per second CFS
cubic inch CU IN.
cubic yard CU YD
current I
current transformer CT
cut out CO
cycles per second CPS

D

damper DMPR
dampproofing DP
dead load DL
decibel DB
deck DK
demolition DML
depth DP
detail DET or DTL
diagonal DIAG
diagram DIAG
diameter D or DIA
dimension DIM.
dimmer DMR
dining room DR
diode DIO
direct current DC
disconnect switch DS
dishwasher DW
distribution DISTR
distribution panel DPNL
division DIV
door DR
dormer DRM
double-acting DBL ACT
double hung window DHW
double-pole DP
double-pole double-throw DPDT
double-pole double-throw switch DPDT SW
double-pole single-throw DPST
double-pole single-throw switch DPST SW
double-pole switch DP SW
double strength glass DSG
double-throw DT
down D or DN
downspout DS
drain DR
drain tile DT
drawing DWG
drinking fountain DF
drum switch DS
dryer D
drywall DW
duplex DX
dust-tight DT
dutch door DD
duty cycle DTY CY
dynamic braking contactor or relay DB

E

each EA
east E
efficiency Eff
ejector pump EP
electric ELEC

. . . ABBREVIATIONS . . .

electrical	ELEC
electrical metallic tubing	EMT
electric panel	EP
electromechanical	ELMCH
electromotive force	EMF
electronic	ELEK
elevation	EL
elevator	ELEV
enamel	ENAM
entrance	ENTR
equipment	EQPT
equipment bonding jumper	EBJ
equipment grounding conductor	EGC
equivalent	EQUIV
estimate	EST
excavate	EXC
exception	Ex.
exhaust	EXH
existing	EXIST. or EXST
expanded metal	EM
expansion joint	EXP JT
explosionproof	EP
exterior	EXT
exterior grade	EXT GR

F

face brick	FB
Fahrenheit	°F
fast	F
field	F
figure	FIG.
fine print note	FPN
finish	FNSH
finish all over	FAO
finished floor	FNSH FL
finished grade	FG or FIN GR
finish one side	F1S
finish two sides	F2S
firebrick	FBCK
fire door	FDR
fire extinguisher	FEXT
fire hydrant	FHY
fireplace	FP
fireproof	FPRF
fire wall	FW
fixed window	FX WDW
fixture	FXTR
flammable	FLMB
flashing	FL
flat	FL
flexible metallic conduit	FMC
float switch	FS
floor	FL
floor drain	FD
flooring	FLG or FLR
floor line	FL
flow switch	FLS
fluorescent	FLUOR or FLUR
flush	FL
flush mount	FLMT
footing	FTG
foot per minute	FPM
foot per second	FPS
foot switch	FTS
forward	F or FWD
foundation	FDN
four-pole	4P
four-pole double-throw switch	4PDT SW
four-pole single-throw switch	4PST SW
four-pole switch	4PSW
frame	FR
frequency	FREQ
front view	FV
full-load amps	FLA
full-load current	FLC
full-load torque	FLT
furnace	FUR.
furring	FUR.
fuse	FU
fuse block	FB
fuse box	FUBX
fuse holder	FUHLR
fusible	FSBL
future	FUT

G

gallon per hour	GPH
gallon per minute	GPM
galvanized	GALV
garage	GAR
gas	G
gate	G
gauge	GA
generator	GEN
glass	GL or GLS
glass block	GLB
glaze	GLZ
gold	Au
grade	GR
grade line	GL
gravel	GVL
gray	GY
green	G or GR
gross vehicle weight	GVW
ground	G, GND, GRD, or GRND
grounded (outlet)	G
ground-fault circuit interrupter	GFCI
ground fault protection of equipment	GFPE
grounding electrode conductor	GEC
grounding electrode system	GES
gypsum	GYP
gypsum sheathing board	GSB

H

hand-off-auto	HOA
handrail	HNDRL
hardboard	HBD
hardware	HDW
hardwood	HDWD
hazardous	HAZ
header	HDR
heating	HTG
heating, air-conditioning, refrigeration	HACR
heating, ventilating, and air conditioning	HVAC
heavy-duty	HD
hertz	Hz
highway	HWY
hollow core	HC
hollow metal door	HMD
horizontal	HOR
horsepower	HP
hose bibb	HB
hot water	HW
hot water heater	HWH
hours	HRS
hydraulic	HYDR

I

immersion detection circuit interrupter	IDCI
inch	IN.
inch per second	IPS
inch-pound	IN. LB
infrared	IR
inside diameter	ID
instantaneous overload	IOL
instantaneous-trip circuit breaker	ITB
insulation	INSUL
integrated circuit	IC
interior	INT or INTR
interlock	INTLK
intermediate	INT
intermediate metal conduit	IMC
International Electrotechnical Commission	IEC
interrupt	INT
inverse time breaker	ITB
inverse-time circuit breaker	ITCB
iron	I
iron pipe	IP
isolated ground	IG

J

jamb	JB or JMB
joint	JT
joist	J or JST
junction	JCT
junction box	JB

K

keyway	KWY
kickplate	KPL
kiln-dried	KD
kilo (1000)	k
1000 circular mils	kcmil
1000′	kFT
kilovolt amps	kVA
kilowatt	kW
kilowatt-hour	kWh
kitchen	K or KIT.
knife switch	KN SW
knockout	KO

L

lamp	LT
lath	LTH
laundry	LAU
laundry tray	LT
lavatory	LAV
left	L
left hand	LH
less-flammable, liquid-insulated	LFLI
library	LBRY or LIB
light	LT
lighting	LTG
lighting panel	LP
lights	LTS
limit switch	LS
line	L
linoleum	LINO or LINOL
lintel	LNTL
live load	LL
living room	LR
load	L
location	LOC
locked-rotor ampacity	LRA
locked-rotor current	LRC
louver	LVR or LV
lumber	LBR

. . . ABBREVIATIONS . . .

M

magnetic brake	MB
main	MN
main bonding jumper	MBJ
main control center	MCC
manhole	MH
manual	MAN., MN, or MNL
manufacturer	MFR
marble	MRB
masonry	MSNRY
masonry opening	MO
material	MATL or MTL
maximum	MAX
maximum working pressure	MWP
mechanical	MECH
medicine cabinet	MC
medium	MED
memory	MEM
metal	MET or MTL
metal anchor	MA
metal door	METD
metal flashing	METF
metal jalousie	METJ
metal lath and plaster	MLP
metal threshold	MT
mezzanine	MEZZ
miles per gallon	MPG
miles per hour	MPH
minimum	MIN
mirror	MIR
miscellaneous	MISC
molding	MLDG
monolithic	ML
mortar	MOR
motor	M, MOT, or MTR
motor branch-circuit, short-circuit, ground-fault	MBCSCGF
motor circuit switch	MCS
motor control center	MCC
motor starter	M
motor switch	MS
mounted	MTD

N

nameplate	NPL
National Electrical Code®	NEC®
National Electrical Manufacturers Association	NEMA
National Electrical Safety Code	NESC
National Fire Protection Association®	NFPA®
negative	NEG
net weight	NTWT
neutral	N or NEUT
nominal	NOM
nonadjustable-trip circuit breaker	NATCB
non-time delay fuse	NTDF
normally closed	NC
normally open	NO
north	N
nosing	NOS
not to scale	NTS
number	NO.

O

Occupational Safety and Health Administration	OSHA
ohmmeter	OHM.
on center	OC
opaque	OPA
opening	OPNG
open web joist	OJ or OWJ
orange	O
ounces per inch	OZ/IN.
outlet	OUT.
outside diameter	OD
overall	OA
overcurrent	OC
overcurrent protection device	OCPD
overhang	OVHG
overhead	OH.
overload	OL
overload relay	OL

P

paint	PNT
painted	PTD
pair	PR
panel	PNL
pantry	PAN.
parallel	PRL
peak-to-peak	P-P
perpendicular	PERP
personal computer	PC
phase	PH
piece	PC
piling	PLG
pillar	PLR
pilot light	PL
piping	PP
pitch	P
plank	PLK
plaster	PLAS
plastic	PLSTC
plate	PL
plate glass	PLGL
plugging switch	PLS
plumbing	PLMB or PLBG
plywood	PLYWD
pneumatic	PNEU
point of beginning	POB
pole	P
polyvinyl chloride	PVC
porcelain	PORC
porch	P
positive	POS
pound(s)	LB
pounds per feet	LB-FT
pounds per inch	LB-IN.
pounds per square foot	PSF
pounds per square inch	PSI
power	P or PWR
power consumed	P
power factor	PF
precast	PRCST
prefabricated	PFB or PREFAB
prefinished	PFN
pressure switch	PS
primary switch	PRI
property line	PL
pull box	PB
pull switch	PS
pull-up torque	PUT
pushbutton	PB

Q

quadrant	QDRNT
quantity	QTY
quarry tile	QT
quarry-tile roof	QTR
quart	QT
quick-acting	QA

R

radius	R
raintight	RT
random	RDM or RNDM
range	R or RNG
receipt of comments	ROC
receipt of proposals	ROP
receptacle	RECEPT or RCPT
recess	REC
recessed	REC
rectifier	REC
red	R
reference	REF
refrigerator	REF
register	REG or RGTR
reinforce	RE
reinforced concrete	RC
reinforcing steel	RST
reinforcing steel bar	REBAR
required	REQD
resistance	R
resistor	R or RES
return	RTN
reverse	R or REV
reverse-acting	RACT
revision	REV
revolutions per minute	RPM
revolutions per second	RPS
rheostat	RH
ribbed	RIB
right	R
right hand	RH
rigid	RGD
riser	R
road	RD
roll roofing	RR
roof	RF
roof drain	RD
roofing	RFG
room	R or RM
root mean square	RMS
rotor	RTR
rough	RGH
rough opening	RO
rough sawn	RS

S

safety switch	SSW
sanitary	SAN
scale	SC
schedule	SCH or SCHED
screen	SCR
scuttle	S
secondary	SEC
section	SECT
selector switch	SS
series	S
service	SERV
service entrance	SE
service factor	SF
sewer	SEW.
shake	SHK
sheathing	SHTHG
sheet	SH or SHT
sheet metal	SM
shelf and rod	SH & RD

. . . ABBREVIATIONS

shelving	SHELV
shingle	SHGL
shower	SH
shutter	SHTR
siding	SDG
silicon controlled rectifier	SCR
sill cock	SC
silver	Ag
single-phase	1PH
single-pole	SP
single-pole circuit breaker	SPCB
single-pole double-throw	SPDT
single-pole double-throw switch	SPDT SW
single-pole single-throw	SPST
single-pole single-throw switch	SPST SW
single-pole switch	SP SW
single strength glass	SSG
sink	S or SK
skylight	SLT
slate	S, SL, or SLT
sliding door	SLD
slope	SLP
slow	S
smoke detector	SD
socket	SOC
soffit	SF
soil pipe	SP
solenoid	SOL
solid core	SC
south	S
spare	SP
specification	SPEC
splash block	SB
square	SQ
square foot	SQ FT
square inch	SQ IN.
square yard	SQ YD
stack	STK
stainless steel	SST
stairs	ST
standard	STD
standpipe	SP
starter	START or STR
steel	STL
stone	STN
storage	STOR
street	ST
structural glass	SG
sump pump	SP
supply	SPLY
surface four sides	S4S
surface one side	S1S
switch	S or SW
switched disconnect	SWD

T

telephone	TEL
television	TV
temperature	TEMP
tempered	TEMP
terazzo	TER
terminal	T or TERM.
terminal board	TB
terra cotta	TC
thermal	THRM
thermally protected	TP
thermostat	THERMO
thermostat switch	THS
three-phase	3PH
three-pole	3P
three-pole double-throw	3PDT
three-pole single-throw	3PST
three-way	3WAY
three-wire	3W
threshold	TH
time	T
time delay	TD
time-delay fuse	TDF
time delay relay	TR
toilet	T
tongue-and-groove	T & G
torque	T
transformer	T, TRANS, or XFMR
transformer, primary side	H
transformer, secondary side	X
tread	TR
triple-pole double-throw	3PDT
triple-pole double-throw switch	3PDT SW
triple-pole single-throw	3PST
triple-pole single-throw switch	3PST SW
triple-pole switch	3P SW
truss	TR
two-phase	2PH
two-pole	DP
two-pole double-throw	DPDT
two-pole single-throw	DPST
typical	TYP

U

unclamp	UCL
underground	UGND
underground feeder	UF
undervoltage	UV
Underwriters Laboratories Inc.	UL
unexcavated	UNEXC
unfinished	UNFIN
up	U
utility room	U RM

V

valley	VAL
valve	V
vapor seal	VS
vaportight	VT
vent	V
ventilation	VENT.
vent pipe	VP
vent stack	VS
vertical	V or VERT
vinyl tile	VTILE or VT
violet	V
volt	V
voltage	E or V
voltage drop	VD
volt amps	VA
volts	V
volts alternating current	VAC
volts direct current	VDC
volume	VOL

W

wainscot	WAIN
walk-in closet	WIC
warm air	WA
washing machine	WM
waste pipe	WP
waste stack	WS
water	WTR
water closet	WC
water heater	WH
water meter	WM
waterproof	WP
watt(s)	W
weatherproof	WP
welded	WLD
welded wire fabric	WWF
west	W
white	W
wide flange	WF
wire gauge	WG
wire mesh	WM
with	W/
without	W/O
wood	WD
wrought iron	WI

Y

yellow	Y

RACEWAYS	
EMT	Electrical Metallic Tubing
ENT	Electrical Nonmetallic Tubing
FMC	Flexible Metal Conduit
FMT	Flexible Metallic Tubing
IMC	Intermediate Metal Conduit
LFMC	Liquidtight Flexible Metal Conduit
LFNC	Liquidtight Flexible Nonmetallic Conduit
RMC	Rigid Metal Conduit
RNC	Rigid Nonmetallic Conduit

COMMON ELECTRICAL INSULATIONS		
60°C 140°F	**75°C 167°F**	**90°C 194°F**
TW UF	FEPW RH RHW THHW THW THWN XHHW USE ZW	TBS SA SIS FEP FEPB MI RHH RHW-2 THHN THHW THW-2 THWN-2 USE-2 XHH XHHW XHHW-2 ZW-2

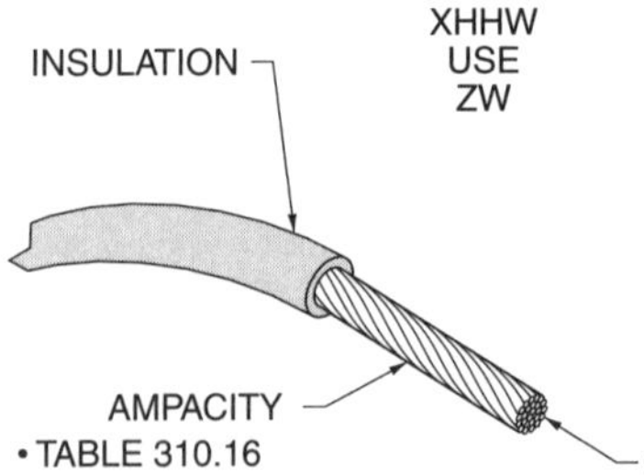

CABLES	
AC	Armored Cable
BX	Tradename for AC
FCC	Flat Conductor Cable
IGS	Integrated Gas Spacer Cable
MC	Metal-Clad Cable
MI	Mineral-Insulated, Metal Sheathed Cable
MV	Medium Voltage
NM	Nonmetallic-Sheathed Cable (dry)
NMC	Nonmetallic-Sheathed Cable (dry or damp)
NMS	Nonmetallic-Sheathed Cable (dry)
SE	Service-Entrance Cable (dry)
TC	Tray Cable
UF	Underground Feeder Cable
USE	Underground Service-Entrance Cable

FUSES AND ITCBs

Increase	Standard Ampere Ratings
5	15, 20, 25, 30, 35, 40, 45
10	50, 60, 70, 80, 90, 100, 110
25	125, 150, 175, 200, 225
50	250, 300, 350, 400, 450
100	500, 600, 700, 800
200	1000, 1200
400	1600, 2000
500	2500
1000	3000, 4000, 5000, 6000

1 A, 3 A, 6 A, 10 A, and 601 A are additional standard ratings for fuses.

HAZARDOUS LOCATIONS

Class	Division	Group	Material
I	1 or 2	A	Acetylene
	1 or 2	B	Hydrogen, butadiene, ethylene oxide, propylene oxide
	1 or 2	C	Carbon monoxide, ether, ethylene, hydrogen sulfide, morpholine, cyclopropane
	1 or 2	D	Gasoline, benzene, butane, propane, alcohol, acetone, ammonia, vinyl chloride
II	1 or 2	E	Metal dusts
	1 or 2	F	Carbon black, coke dust, coal
	1 or 2	G	Grain dust, flour, starch, sugar, plastics
III	1 or 2	No Groups	Wood chips, cotton, flax, nylon

Glossary

A

accessible: Capable of being reached for operation, renewal, and inspection.

alternating current (AC): Current that reverses its direction of flow at regular intervals.

ambient temperature: The temperature of air around a conductor or a piece of equipment.

ampacity: The maximum current, in amperes, that a conductor can carry continuously under the conditions of use without exceeding its temperature rating.

apparent power: The product of voltage and current in a circuit calculated without considering the phase shift that may be present between total voltage and current in the circuit.

approved: Acceptable to the authority having jurisdiction (AHJ).

attachment plug (plug cap or plug): A device that, by insertion in a receptacle, establishes a connection between the conductors of the attached flexible cord and the conductors connected permanently to the receptacle.

B

bare conductor: A conductor having no covering or electrical insulation whatsoever.

bonded (bonding): Connected to establish electrical continuity and conductivity.

branch circuit: The circuit conductors between the final overcurrent device protecting the circuit and the outlet(s).

building: A structure that stands alone or that is separated from adjoining structures by fire walls.

C

capacitance: The ability of a component or circuit to store energy in the form of an electrical charge.

combustible gas detection system: A protection technique utilizing stationary gas detectors in industrial establishments.

compact conductor: A conductor that has been compressed to eliminate voids between strands.

continuous duty: Operation at a substantially constant load for an indefinitely long time.

continuous load: A load where the maximum current is expected to continue for 3 hours or more.

covered conductor: A conductor encased within material of composition or thickness that is not recognized by the NEC® as electrical insulation.

current: The flow rate of electrons in a circuit.

D

device: A unit of an electrical system, other than a conductor, that carries or controls electrical energy as its principal function.

direct current (DC): Current that flows in one direction.

disconnecting means: A device or group of devices that separate or isolate the conductors of a circuit from their source of supply.

dwelling unit: A single unit, providing complete and independent living facilities for one or more persons, including permanent provisions for living, sleeping, cooking, and sanitation.

E

eddy current: An unwanted, induced current in the core of a transformer.

efficiency: The output divided by the input of a motor.

electric sign: A fixed, stationary, or portable self-contained, electrically operated and/or electrically illuminated utilization equipment with words or symbols designed to convey information or attract attention.

energized: Electrically connected to, or is, a source of voltage.

equipment grounding conductor (EGC): A conductive path(s) that is part of an effective ground-fault current path and connects normally non-current-carrying metal parts of equipment together and to the system grounded conductor or to the grounding electrode conductor, or both.

exposed (as applied to live parts): Capable of being inadvertently touched or approached nearer than a safe distance by a person.

exposed (as applied to wiring methods): On or attached to the surface or behind panels designed to allow access.

F

feeder: All circuit conductors between the service equipment, the source of a separately derived system, or other power supply source and the final branch-circuit overcurrent device.

fitting: An accessory such as a locknut, bushing, or other part of a wiring system that is intended primarily to perform a mechanical rather than an electrical function.

G

grounded (grounding): Connected (connecting) to ground or to a conductive body that extends the ground connection.

grounded conductor: A system or circuit conductor that is intentionally grounded.

guarded: Covered, shielded, fenced, enclosed, or otherwise protected by means of suitable covers, casings, barriers, rails, screens, mats, or platforms to remove the likelihood of approach or contact by persons or objects to a point of danger.

H

handhole enclosure: An enclosure for use in underground systems, provided with an open or closed bottom, and sized to allow personnel to reach into, but not enter, for the purpose of installing, operating, or maintaining equipment or wiring or both.

hoistway: Any shaftway, hatchway, well hole, or other vertical opening or space in which an elevator or dumbwaiter is designed to operate.

hysteresis: A lagging in values resulting in a changing magnetization in a magnetic material.

I

in sight from (within sight from, within sight): Where the NEC® specifies that one equipment shall be "in sight from," "within sight from," or "within sight of," and so forth, another equipment, the specified equipment is to be visible and not more than 15 m (50′) distant from the other.

insulated conductor: A conductor encased within material of composition and thickness that is recognized by the NEC® as electrical insulation.

interrupting rating: The highest current at rated voltage that a device is identified to interrupt under standard test conditions.

K

K factor: The resistance of a circular-mil foot of wire at a set temperature.

L

labeled: Equipment or materials to which has been attached a label, symbol, or other identifying mark of an organization that is acceptable to the authority having jurisdiction (AHJ) and concerned with product evaluation, that maintains periodic inspection of production of labeled equipment or materials, and by whose labeling the manufacturer indicates compliance with appropriate standards or performance in a specified manner.

live parts: Energized conductive components.

loss: The difference between a motor's input and output.

M

main bonding jumper: The connection between the grounded circuit conductor and the equipment grounding conductor (EGC), or the supply-side bonding jumper, or both, at the service.

multiple fuse: An assembly of two or more single-pole fuses.

mutual induction: Voltage caused in one circuit by a change in current of another circuit.

N

nominal voltage: A nominal value assigned to a circuit or system for the purpose of conveniently designating its voltage class (e.g. 120/240 V, 480Y/277 V, 600 V).

nonlinear load: A load where the wave shape of the steady-state current does not follow the wave shape of the applied voltage.

O

Ohm's law: The relationship between the voltage measured in volts (V), current measured in amperes (A), and resistance measured in ohms (Ω) in an electrical circuit.

oil immersion: Electrical equipment immersed in a protective liquid in such a way that an explosive atmosphere that may be above the liquid or outside the enclosure cannot be ignited.

overcurrent: Any current in excess of the rated current of equipment or the ampacity of a conductor. It may result from overload, short circuit, or ground fault.

overloads: Heat-sensing devices intended to protect the motor from excessive current.

P

parallel circuit: A circuit with two or more paths for current to flow.

power factor (PF): The electrical efficiency of a circuit determined by the ratio of true power used in an AC circuit to apparent power delivered to the circuit.

Q

qualified person: One who has skills and knowledge related to the construction and operation of electrical equipment and installations and has received safety training to recognize and avoid the hazards involved.

R

raintight: Constructed or protected so that exposure to a beating rain will not result in the entrance of water under specified test conditions.

ratio: The whole numbers designating the relationship between high voltage and low voltage.

receptacle outlet: An outlet where one or more receptacles are installed.

reciprocal: The inverse relationship of two numbers.

resistance: The opposition to current flow.

S

series circuit: A circuit with only one path for current to flow.

service drop: The overhead conductors between the serving utility and the service point.

service lateral: The underground conductors between the utility electric supply system and the service point.

service point: The point of connection between the facilities of the serving utility and the premises wiring.

show window: Any window, including windows above doors, used or designed to be used for the display of goods or advertising material, whether it is fully or partly enclosed or entirely open at the rear and whether or not it has a platform raised higher than the street floor level.

special permission: The written consent of the authority having jurisdiction (AHJ).

substation: An assemblage of equipment (e.g., switches, interrupting devices, circuit breakers, buses, and transformers) through which electric energy is passed for the purpose of distribution, switching, or modifying its characteristics.

T

temperature: A measurement of the intensity of heat.

temperature rise: The amount of heat that an electrical component produces above the ambient temperature.

thermal protector (as applied to motors): A protective device for assembly as an integral part of a motor or motor-compressor that, when properly applied, protects the motor against dangerous overheating due to overload and failure to start.

true power: The actual work being done by the load using electricity.

V

voltage: The amount of electrical pressure in a circuit.

voltage drop: The voltage that is lost due to the resistance of conductors.

W

watertight: Constructed so that moisture will not enter the enclosure under specified test conditions.

Index

Page numbers in italic refer to figures.